Zirconium Alloy Coatings

Junhua Hu · Guoqin Cao

Zirconium Alloy Coatings

Advances in Research and Applications

Springer

Junhua Hu
School of Materials Science
and Engineering
Zhengzhou University
Zhengzhou, China

Guoqin Cao
School of Materials Science
and Engineering
Zhengzhou University
Zhengzhou, China

ISBN 978-981-96-5027-9 ISBN 978-981-96-5028-6 (eBook)
https://doi.org/10.1007/978-981-96-5028-6

This Springer imprint is published by the registered company Springer Nature Singapore Pte Ltd.
The registered company address is: 152 Beach Road, #21-01/04 Gateway East, Singapore 189721, Singapore

Preface

Zirconium alloys are widely used in nuclear fuel cladding due to their low thermal neutron absorption cross-section, high thermal conductivity, and good mechanical properties. However, zirconium alloys are prone to oxidation in high-temperature water environments, leading to serious corrosion issues. Traditional metallic coatings can provide short-term protection but face challenges in interface reactions and element migration, making it difficult to maintain structural integrity over the long term. This comprehensive book delves into the realm of Zr-Si-based coatings, exploring their innovations in research methodologies, and pivotal findings, covering the multidisciplinary field of materials science and surface engineering. This book also offers a detailed analysis of the development, properties, and application prospects of zirconium alloy coatings. Readers will find compelling insights into cutting-edge research methods, innovative findings, and emerging trends in the field.

The book elucidates the fundamental principles governing zirconium alloy coatings, highlighting their role in corrosion resistance, high-temperature oxidation resistance, and industrial applications. Notably, it presents a rich array of illustrative figures and tables, enhancing comprehension and facilitating a deeper understanding of the subject matter. The theoretical calculation and advanced structural characterization are considered comprehensively to ensure the reliability of service performance of the coating.

This book caters to a wide spectrum of readers, including researchers, engineers, and practitioners in materials science, surface engineering, and metallurgy. It serves as a valuable reference for academia, providing a comprehensive understanding of zirconium alloy coatings and their practical implications across various industries, from aerospace to marine applications.

Zhengzhou, China

Junhua Hu
Guoqin Cao

Contents

Chapter 1
Development of Nuclear Reactors and New Requirements for Cladding Materials

Abstract Nuclear energy plays a crucial role in addressing global energy and environmental challenges, offering a low-carbon, efficient, and sustainable power source. The development of nuclear reactors has evolved through four generations, focusing on improving safety, performance, and sustainability, with pressurized water reactors dominating the global market. A key component of nuclear reactor safety is the fuel cladding, which prevents radioactive leakage and must exhibit exceptional mechanical properties, corrosion resistance, and minimal neutron absorption. Zirconium-based alloys have been the material of choice for decades, with advancements like Zr-2 and Zr-4 improving their performance. However, during severe accidents, such as those at Fukushima, zirconium reacts with steam, generating hydrogen and increasing risk. Ongoing research seeks to enhance cladding materials, focusing on improving their performance under extreme conditions to ensure nuclear plant safety and operational efficiency.

Keywords Nuclear energy · Cladding materials · Zirconium alloy · Accident-tolerant fuel cladding

1.1 Development of Nuclear Energy

In the evolution of human civilization, energy is not only the core driving force for global development and economic growth, but also constitutes the material cornerstone of human society. With the gradual recovery of the global economy in the later stages of the epidemic, the demand for energy has shown an increasing trend. Total global electricity generation has continued to climb in recent years. However, thermal power generation, which relies on the consumption of large amounts of non-renewable energy resources, still dominates, and this dependence poses a constant energy threat to countries around the world. It is predicted that oil and gas reserves support production for only the next 50 years or so, while coal reserves will last only about 100 years.

J. Hu and G. Cao, *Zirconium Alloy Coatings*,
https://doi.org/10.1007/978-981-96-5028-6_1

The extensive exploitation and consumption of fossil fuels not only pose the risk of depletion of energy resources, but also trigger a series of serious social and environmental problems, including global warming, air quality degradation, acid rain, etc., which collectively constitute a major threat to the living environment of mankind and to the sustainable development of the economy. Among these problems, global warming is particularly prominent and has become a major challenge facing the global community at present. Observational data show that the average temperature of the earth has risen by about 0.87 °C during the decade 2006–2015, when industrialization began. Currently, the Earth's surface temperature is increasing at a rate of about 0.2 °C per decade. Continuing to rise at this rate, the Earth's average temperature is projected to be 1.5 °C above pre-industrial levels by 2037 [1].

This global temperature rise will have far-reaching impacts on the Earth's climate system, leading to frequent extreme weather events such as high temperatures, typhoons, and blizzards, which not only threaten the health and well-being of human societies, but also pose a great challenge to ecosystems and biodiversity. Therefore, mitigating global warming, reducing greenhouse gas emissions, and transitioning to sustainable energy use have become urgent global tasks. From a long-term perspective, ensuring a sustainable supply of energy is not only the basis for achieving economic growth, but also the key to maintaining the global ecological balance and promoting the long-term prosperity of human society. Against this background, accelerating the development of renewable energy technologies, improving energy efficiency, and promoting the transformation and upgrading of the energy structure have become the direction of joint efforts by governments and international organizations. Among them, nuclear energy is regarded as a very important source of electricity because of its low environmental impact, low-resource consumption, low-power generation costs, high specific energy, and low-carbon emissions. Nuclear power accounts for less than one sixth of total thermal power generation and, as the main hope for the development of nuclear power, the current share of nuclear energy in emerging developing countries is still relatively low, indicating that nuclear energy still has great potential in global energy development.

The world's first nuclear reactor was built at the University of Chicago in December 1942, proving that people could control the nuclear fission chain reaction. Due to the era of World War II, mankind took the lead in applying this new technology to warfare. Afterward, major countries gradually turned military nuclear power reactor technology to civilian use, inaugurating of nuclear power construction. Generally speaking, the development of the world's nuclear power technology can be divided into four generations, as shown in Fig. 1.1.

First-Generation Nuclear Power Technology: In the 1950s and 1960s, based on military nuclear reactor technology, the first batch of pilot reactors or prototype power plant were designed, developed and constructed by the USA, the Soviet Union, Canada, the UK, and other countries, which verified the technical feasibility of nuclear power generation. After the technical feasibility of first-generation nuclear energy systems was verified, the nuclear power industry implemented standardization, serialization, and batch construction of these verified models in the 1970s to 1990s, and the vast majority of nuclear power plants that are still in commercial

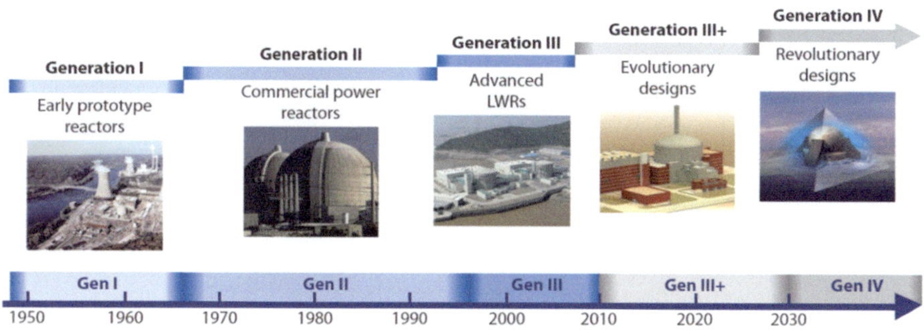

Fig. 1.1 Generations of nuclear power: Time ranges correspond to the design and the first deployments of different generations of reactors

operation belong to second-generation or second-generation improved technologies. This period was a time of great development of commercial nuclear power plants, and several major types of nuclear power units in different countries during this period are listed in Table 1.1.

The design of Generation III reactors builds upon the fundamental principles of Generation II reactors, incorporating decades of operational insights gained from these reactors. It further leverages newly developed technologies that have undergone rigorous validation and demonstration of feasibility. The overarching objectives are to enhance the safety of existing reactors and to comply with the requirements stipulated in the US Utility Requirements Document (URD) and the European Utility Requirements (EUR). The development of Generation III nuclear power systems began in the 1990s, with the main models being the AP1000, EPR, ABWR, APR1400, AES2006, ESBWR, CAP1400, Hualong-1, and Guohe-1.

Generation IV nuclear power technology is a new generation of advanced nuclear energy systems for the future, with major innovations and developments in both reactors and fuel cycles. The development goals of Generation IV nuclear power systems are to enhance energy sustainability, the economic competitiveness, safety and reliability of nuclear power plants, and proliferation resistance and physical protection. The Generation IV International Forum (GIF) was established in 2001 to conduct research on and develop Generation IV nuclear energy systems. It currently has 13

Table 1.1 Main types of second-generation nuclear power units in different countries

Design country	Nuclear power unit type
USA	Pressurized water reactor nuclear power type (PWR, System 80)
	Boiling water reactor (BWR)
France	Pressurized Water Reactor (P4, M310)
Russia	Light-water reactor (VVER)
Canada	Heavy-water reactor nuclear power models (CANDU)

member countries as well as Euratom, which represents the 28 countries of the European Union. The focus is on six advanced nuclear energy systems, including sodium-cooled fast reactor (SFR), very high-temperature gas-cooled reactor (VHTR), supercritical water reactor (SCWR), gas-cooled fast reactor (GFR), molten salt reactor (MSR), and lead-cooled fast reactor (LFR) [2].

Until today, nuclear energy as an efficient, green energy source has been widely used around the world. According to the International Atomic Energy Agency data statistics, as of the end of 2023, the total installed capacity of the world's 412 operating nuclear power plants is 370,170 MW, and the world's total amount of nuclear power generation is more than 2653 TWh, which is about 10% of the world's total power generation, and more than 25% of the world's low-carbon power generation, and nowadays, the development of nuclear power has become one of the important strategic plans of various countries. As of 2023, pressurized water reactors (PWRs) are the most prevalent type of nuclear reactor in operation globally, accounting for 2/3 of all nuclear power plants, closely followed by boiling water reactors (BWRs), which account for 1/6 of the reactors in operation globally, and pressurized heavy water reactors (PHWRs), which account for about 1/10 of all reactors in operation. These are the main types of reactors of nuclear power plants currently in operation.

Table 1.2 reflects the current status of nuclear reactor development globally (as of 2023, data from IAEA's Country Nuclear Profiles). Among these countries, the USA, France, China, Russia, and South Korea are substantially ahead of other countries in terms of nuclear installed capacity and electricity generation.

1.2 Nuclear Reactor Cladding Materials

In the development and operation of nuclear power plants, safety is always the most critical consideration. From site selection to design, to construction and daily operation, the principle of prioritizing safety should be strictly followed in all aspects. As the core component of a nuclear power plant, the design and construction of nuclear reactors require special attention to ensure the safety and stability of the entire plant.

Taking a pressurized water reactor as an example, the composition of its nuclear reactor pressure vessel includes a number of key parts such as the core, the reflector layer, the control elements, and the reactor vessel. The core is the key area where the chain nuclear reaction takes place. It consists of a number of components such as the nuclear fuel assembly, the control assembly, the combustible poison assembly, and the neutron source assembly and is filled with water as the moderator and coolant between these components. A typical core may contain one to two hundred nuclear fuel assemblies, each of which in turn consists of more than two hundred nuclear fuel elements, and each of which is made of several hundred core blocks of uranium dioxide (UO_2) nuclear fuel packed into alloy cladding tubes, as shown in the schematic diagram of the nuclear fuel assembly in Fig. 1.2. Therefore, alloy cladding tubes are in great demand for pressurized water reactor nuclear power plants [3].

Table 1.2 Development status of global nuclear reactors

Country	Number of nuclear reactors in operation	Total installed capacity (MWe)	Total annual electricity generation (GW(e)·h)	Nuclear's share of the country's total electricity generation (%)
USA	93	95,835	772,220	18
France	56	63,370	279,000	63
China	55	53,181	407,141	5
Russia	37	27,727	223,372	20
South Korea	25	24,489	167,346	30
Canada	19	13,624	82,301	13
Ukraine	15	13,107	83,003	55
Japan	12	11,046	51,774	6
Spain	7	7123	55,599	20
Sweden	6	6937	50,063	29
India	19	6290	46,195	3
UK	9	5883	45,868	14
Finland	5	4394	24,221	35
Czech Republic	6	3934	31,021	37
Belgium	5	3928	41,617	46

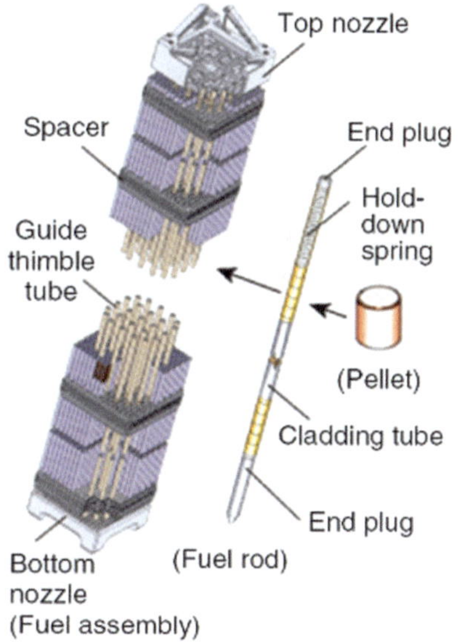

Fig. 1.2 Diagram of nuclear fuel assembly [5]

During the operation of nuclear power plants, it is extremely important to prevent the leakage of radioactive elements into the environment. For this reason, nuclear power plants are designed with multiple barriers to ensure the isolation of radioactive elements from the external environment. The fuel cladding plays a crucial role as the first barrier. While it encapsulates and seals the uranium oxide core block, it also serves to prevent the release of radioactive elements. Therefore, the selection of fuel cladding materials and their performance are of utmost criticality. Such claddings must exhibit excellent mechanical properties—capable of withstanding coolant erosion, vibration, and additional stresses induced by residual stresses, thermal cycling, and fuel expansion, among others. They must also demonstrate robust corrosion and radiation resistance, coupled with a neutron capture cross-section as small as feasible to minimize neutron loss from nuclear fuel reactions. These attributes collectively ensure stable and reliable operation under extreme service conditions. This requires in-depth research and strict control of the alloy cladding in many fields, such as materials engineering, nuclear engineering and safety design, in order to meet the high standards of safe operation of nuclear power plants [4].

Table 1.3 lists some of the potential substitute cladding materials with their neutron absorption cross-sections and yield strength values. Be's excellent performance, its poor ductility and insufficient corrosion resistance, and tendency to produce toxic reactions make it unsuitable for various applications, leading to its discard. Zr, with its low thermal neutron capture cross-section and its alloying properties, stood out among the candidates. The development of industrial zirconium-based alloys was initiated early in the design of water-cooled reactors and has continued.

The development of industrial zirconium-based alloys to improve the neutronic efficiency of water-cooled reactors began as early as the design of nuclear reactors

Table 1.3 Neutron absorption cross-section and yield strength of certain materials

Elements	Neutron absorption cross-section (barns)	Yield strength (MPa)
Be	0.009	200–350
C	0.004	24–28
Mg	0.063	65–100
Si	0.160	165–180
Zr	0.185	135–310
Al	0.231	30–40
Mo	2.480	170–350
Cr	3.050	185- -280
Nb	1.150	75–95
Fe	2.550	110–165
Ni	4.430	8–280
V	5.040	125–180
Sn	0.630	7–15

and has continued, with the first use of Zr in fuel cladding for US Navy submarine light-water reactors (LWRs) in 1949, and all existing cladding materials in light-water reactors are being replaced by Zr-based alloys in the mid-1960s, and with the advent of the Canadian-designed natural uranium-fueled, pressurized heavy water reactors (PHWRs), the use of Zr-based alloys as structural materials for nuclear reactors was further developed [6].

However, the corrosion resistance of Zr in water and steam environments is not satisfactory and alloying is needed to improve corrosion resistance under reactor conditions. Researchers developed the Zr-1 alloy, which adds 2.5 wt.% Sn to Zr. The addition of Sn primarily mitigates the detrimental effects of N and, to some extent, C and Al, but it was found that Sn reduces the overall corrosion resistance of Zr. On this basis, the Zr-2 alloy was developed, to which only 1.5 wt.% of Sn was added to the Zr-1 alloy, while the contents of Fe, Cr, and Ni were limited to 0.12 wt.%, 0.1 wt.%, and 0.05 wt.%, respectively. The corrosion resistance of the Zr-2 alloy in high-temperature aqueous environments was superior to that of the unalloyed Zr. In order to further improve the performance of the Zr-2 alloy, in the mid-1950s, a new alloy was developed in the mid-1950s. In order to further improve the performance of the Zr-2 alloy, in the mid-1950s, a new alloy, the Zr-3 alloy, was developed. However, the localized appearance of white corrosion deposits oriented in the form of thin stripes in the rolling direction of Zr-3 alloy strips raised concerns about the performance of the alloy and led to its abandonment. At the same time, the researchers found that Ni-bearing zirconium alloys were susceptible to hydrogen uptake corrosion. After replacing Ni with Fe in Zr-2 alloy, researchers proposed Zr-4 alloy, which was widely promoted for its low hydrogen absorption and good corrosion resistance, and was used in a large number of pressurized water reactors (PWRs) for cladding materials to reduce the absorption of hydrogen in cladding materials from the light-water-coolant [6]. In the former Soviet Union, E110, E110G, and E125 binary Zr–Nb alloys were developed with Nb contents of 1 wt.% and 2.5 wt.%, and their doping can improve the corrosion resistance of the alloys [7, 8].

Despite the proven performance of Zr–Nb alloys, Zr-2 and Zr-4 alloys, researchers continued to develop Valloy and Valloy S alloys with excellent cladding properties. The Soviet Union also developed multicomponent α-Ozhenite alloys containing Sn, Fe, Ni, and Nb in the 1960s and E635 alloy in the 1970s. The German Metallgesellschaft AG (Metallgesellschaft AG) developed Zr-3%Nb-1%Sn alloys that are stronger than zirconium alloys [9]. ATR alloys for CO^{2-} cooled reactors were developed in the UK. Westinghouse developed ZIRLO alloy which has lower irradiation-induced creep and growth rates and better corrosion resistance compared to Zr-4 [10]. They have also developed the AXIOM alloy for use in more challenging environments [11]. Mitsubishi Heavy Industries (MHIs) and Nuclear Fuel Industries (NFIs) developed MDA alloys with better corrosion resistance and lower creep and growth rates than low Sn zirconium alloys. NFI and Sumitomo Metals Industries (SMI) developed NDA alloys with reduced Nb and Sn content and increased Fe and Cr content, which improves the alloys' resistance to off-of-pile corrosion [12]. Japan developed J series alloys, and Korea also conducted experiments on new multicomponent Zr-based cladding alloys to develop HANA alloys [13].

Siemens developed a new generation of dual-phase cladding material for pressurized water reactors, called DX ELS 0.8, with an outer layer consisting of a 100 mm-modified Zr-4 alloy with lower Sn content and increased Fe and Cr and an inner layer of standard Zr-4 alloy. This material not only has high corrosion resistance, but also inherits the mechanical strength of the Zr-4 alloy under normal and transient operating conditions, as well as assumed Loss-of-Coolant Accident (LOCA) conditions. Around 2010, it was replaced by another DX cladding tube with better corrosion resistance, the DX-D4 (with 0.5% Sn). In the early 1990s, Framatome developed new cladding materials with improved properties, including ultra-low Sn zirconium alloys such as M3, M4, and M5 [14].

These zirconium alloys offer a significant improvement in corrosion resistance compared to conventional alloys, as well as machinability, but corrosion and hydrogen adsorption in aqueous environments of nuclear fuel cladding materials remain potential limiting factors for the operation of nuclear fuel at high burnups and under more extreme conditions. In 1979, a lack of cooling water for the core of the Three Mile Island Nuclear Power Plant (TMISP) in the US Peninsular States resulted in a dramatic increase in temperature over a short period of time, which led to a core. In 1986, the Chernobyl nuclear power plant in the USSR overheated the core due to operational errors by the staff, causing the zirconium alloy cladding tubes to react violently with water and the resulting hydrogen to explode at ultra-high temperatures, an accident that posed considerable hazards. In 2011, the Fukushima nuclear power plant in Japan, due to the earthquake and tsunami, among other things, lost power supply and the cooling water cycle was stopped. And the excessive accumulation of fission-generated heat led to a rapid reaction of the zirconium alloy cladding with overheated water and water vapor, as shown in Eq. (1.1):

$$2H_2O\ (l, g) + Zr(s) = ZrO_2(s) + 2H_2(g). \tag{1.1}$$

A large amount of H_2 is generated and accumulated, while a large amount of heat is generated due to the exothermic oxidation reaction of Zr, leading to an explosion.

In severe accidents, core cooling is usually interrupted, and decay heat in the fuel raises the core temperature rapidly. As the amount of water in the pile decreases, heat transfer efficiency rapidly decreases, and the fuel elements begin to undergo various physical and chemical degradations at high temperatures. Physical degradation occurs at 700–1000 °C and involves expansion and rupture of the cladding tubes. Chemical degradation is mainly determined by steam oxidation of zirconium alloys. For example, about 125 kg of zirconium metal per PWR fuel assembly is oxidized to produce some hydrogen. If the zirconium metal was fully oxidized, the additional heat generated would aggravate the accident severity and further increase the burden on the emergency core cooling system (ECCS).

Studies have demonstrated that the mechanical properties and oxidation resistance of alloys are severely degraded in high-temperature environments, especially above 800 °C. During conventional oxidation, a black layer of monoclinic phase ZrO_2 (m-ZrO_2) forms on the surface of the Zr alloy. This dense oxide layer maintains its stability and good corrosion resistance in a variety of environments, including acidic, alkaline, and aqueous media at high temperature and pressure. This oxide layer provides effective protection for the Zr alloy and extends its service life. When the temperature rises above 1173 °C, ZrO_2 undergoes a phase transition from a monoclinic phase to a tetragonal phase ZrO_2 (t-ZrO_2). This phase transition is accompanied by an increase in volume and the generation of internal stresses that lead to stress concentrations regions in the oxide film, which tend to initiate cracks in the oxide film, especially in the regions close to the substrate, where the formation of transverse cracks becomes particularly pronounced. As the oxidation process proceeds, these transverse cracks will gradually expand and increase in size, which may eventually lead to the peeling of the oxide film, causing the Zr alloy to lose its protective layer and be exposed to the corrosive environment, thus accelerating the corrosion process [15].

The fission of heavy nuclei of atoms can release huge amounts of energy and can cause considerable harm to the environment. Engineers designing nuclear reactors have proposed strategies to address the needs of nuclear safety systems, i.e., assuming that two extreme boundary accidents are assumed, namely, Reactivity Insertion Accident (RIA) and Loss-of-Coolant Accident (LOCA), and an active safety system (Emergency Core Cooling System, ECCS) has been designed to cope with these assumed Design Basis Accidents (DBAs) by maintaining a controllable and coolable core. Table refers to these accidents, which are highly unlikely to occur but can have significant and severe consequences if they do, as black swan events [16]. Some researchers have predicted that the probability of estimating these black swan events is extremely small, approximately less than 10–5 reactor-years [17], and therefore, researchers have not given much consideration to the design basis of ECCS. However, the reality is that the rate of nuclear power accidents is much higher than the predicted value, and the actual rate is higher than 2×10^{-4}, which is a serious problem for the nuclear safety field that cannot be ignored.

Therefore, the reliability of Zr alloys under extreme conditions, especially under severe accident conditions such as LOCA, needs to be managed and controlled through a combination of material improvements, engineering safety precautions, and emergency response strategies to ensure the safe operation of nuclear power plants. As a result, the nuclear industry has proposed the idea of replacing Zr alloy cladding with accident-tolerant fuel (ATF) cladding, and the development of ATF cladding materials focuses on the exploration materials that can maintain better mechanical stability and lower oxidation rates at high temperatures and on the enhancement of the accident-tolerant capability of nuclear cladding materials to reduce the damage to the fuel cladding in the event of severe accidents such as loss of water accidents (LOCA). By increasing the fault tolerance capability of the nuclear cladding, the oxidation rate and heat release from the fuel cladding can be reduced in the event of a severe accident, such as a water loss accident, delaying the core meltdown and effectively suppressing the amount of hydrogen generated, providing an additional window of time to take measures to respond to the accident, as shown in Fig. 1.3.

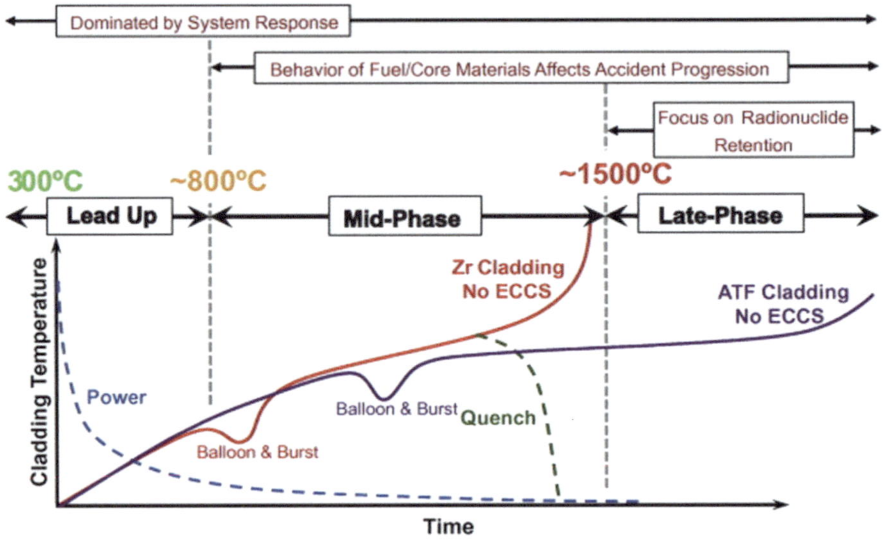

Fig. 1.3 Schematic diagram of accident-tolerant fuel cladding for accident deferral [18]

1.3 Conclusion

This chapter outlines the progression of nuclear energy, its significance in tackling worldwide energy issues, and the crucial role of fuel cladding in the security and performance of nuclear reactors. As a clean and high-energy–density source of power, nuclear energy has become a key part of the global energy structure. With technological advancements, nuclear reactors have evolved to the fourth generation, focusing on sustainability and safety. Zirconium alloys, used as fuel cladding materials, face challenges in oxidation, hydrogen absorption, and mechanical stability at high temperatures, which could threaten the safety of the reactor. To address these challenges, the development of accident-tolerant fuel (ATF) cladding materials has become a priority, aiming to enhance oxidation resistance and mechanical properties at high temperatures, reduce hydrogen generation and heat release during severe accidents, lower the risk of core meltdown, and enhance the safety of nuclear power plants. This chapter emphasizes the importance of ongoing research into advanced cladding materials to support the long-term development of nuclear energy as a sustainable and safe source of power.

References

1. Liang S, Wang D, Ziegler AD et al (2022) Madden–Julian oscillation-induced extreme rainfalls constrained by global warming mitigation. NPJ Clim Atmos Sci 5(1):67
2. Pioro I (2017) Handbook of generation-IV nuclear reactors. J Nucl Eng Radiat Sci 3(2)

3. Rebak RB (2017) Iron-chrome-aluminum alloy cladding for increasing safety in nuclear power plants. EPJ Nuclear Sci Technol 3:34
4. Ming Z, Yingxin L, Shaojie O et al (2016) Nuclear energy in the Post-Fukushima era: research on the developments of the Chinese and worldwide nuclear power industries. Renew Sustain Energy Rev 58:147–156
5. Tang C, Stueber M, Seifert HJ et al (2017) Protective coatings on zirconium-based alloys as accident-tolerant fuel (ATF) claddings. Corros Rev 35(3):141–165
6. Tewari R, Krishna KVM, Neogy S et al (2020) Zirconium and its alloys: properties and characteristics pp 284–302
7. Király M, Hózer Z, Horváth M et al (2019) Impact of thermal and chemical treatment on the mechanical properties of E110 and E110G cladding tubes. Nucl Eng Technol 51(2):518–525
8. Hózer Z, Perez-Feró E, Novotny T et al (2015) Experimental comparison of the behavior of E110 and E110G claddings at high temperature. ASTM Spec Tech Publ 932–951
9. Anderko K, Jung-König W, Richter H et al (1962) Beitrag zur Entwicklung korrosionsbeständiger Zirkoniumlegierungen mit hoher Festigkeit für den Reaktorbau 53(8):503–512
10. Hudson D, Smith GDW (2009) Initial observation of grain boundary solute segregation in a zirconium alloy (ZIRLO) by three-dimensional atom probe. Scripta Mater 61(4):411–414
11. Pan G, Long CJ, Garde AM et al (2010) Advanced material for PWR application: AXIOM™ cladding. LWR Fuel Perf Meet/Top Fuel/WRFPM 2010:612–620
12. Isobe T, Matsuo Y (1991) Development of highly corrosion resistant zirconium-base alloys
13. Song KW, Jeong YH, Kim KS et al (2008) High burnup fuel technology in Korea. Nucl Eng Technol 40(1):21
14. Steinbrück M, Vér N, Große M (2011) Oxidation of advanced zirconium cladding alloys in steam at temperatures in the range of 600–1200 °C. Oxid Met 76(3):215–232
15. Cox B (2005) Some thoughts on the mechanisms of in-reactor corrosion of zirconium alloys. J Nucl Mater 336(2–3):331–368
16. Taleb NN (2007) The black swan: the impact of the highly improbable. Random house
17. D'Auria FS, Dusic M, Dutton LMC et al (2009) Deterministic safety analysis for nuclear power plants. IAEA, IAEA Specific Safety Guide
18. Zinkle SJ, Terrani KA, Gehin JC et al (2014) Accident tolerant fuels for LWRs: a perspective. J Nucl Mater 448(1–3):374–379

Chapter 2
Research Progress and Performance evaluation of Zirconium Alloy Coatings

Abstract To enhance the oxidation resistance of zirconium alloy fuel cladding in high-temperature steam environments, current research primarily focuses on improving its durability and safety through surface coating technologies. This paper reviews various coating systems applied to zirconium alloys, including Cr coatings, FeCrAl coatings, high-entropy coatings, ceramic coatings, and MAX-phase coatings. These coating technologies effectively mitigate oxidation reactions by adding a protective layer on the zirconium alloy surface, thereby slowing down the oxidation rate of zirconium alloys under extreme accident scenarios, such as Loss-of-Coolant Accidents (LOCA), and improving the safety of nuclear reactors under harsh conditions. However, the long-term stability and high-temperature limits of these coatings remain key areas of research. Further optimization of coating materials' composition and fabrication processes is needed to enable broader industrial applications.

Keywords Zirconium alloy · High-temperature vapor · Cr coating · FeCrAl coating · High-entropy coating · Ceramic coating · MAX-phase coating

2.1 Zirconium Alloy Protective Coating

In the previous chapter, it was mentioned that enhancing the capability of the Zr alloy fuel cladding to resist high-temperature vapor oxidation is an important research direction in the development of ATF nuclear fuel technology. This is because under extreme conditions such as LOCA, the oxidation of core materials is mainly caused by the oxidation of Zr alloy cladding. Therefore, optimization of Zr-based cladding materials to improve their oxidation resistance under high-temperature vapor environments is essential to enhance the safety of nuclear reactors.

For the high-temperature steam oxidation resistance of ATF fuel cladding, the industry currently employs two main strategies to enhance the performance of cladding materials. First, exploration and development of novel antioxidant materials are one path. These include SiCf/SiC composites, FeCrAl alloys, Ti–Al alloys, and Mo materials, which are considered as promising alternative options due to their

J. Hu and G. Cao, *Zirconium Alloy Coatings*,
https://doi.org/10.1007/978-981-96-5028-6_2

excellent antioxidant properties [1]. However, the introduction of these new material systems may be accompanied by specific challenges, such as the aqueous solubility of SiC in water and the complexity of fabrication and molding, which may lead to a longer research and development cycle and higher costs for new fuel cladding. In addition, the adoption of these new cladding materials may require significant engineering redesign of the nuclear reactor core, representing a long-term development strategy. An alternative response strategy is to enhance the oxidation resistance of Zr alloy cladding currently in use through surface modification techniques, particularly the application of protective coatings. This approach, by adding micron-level coatings to the surface of Zr alloys, does not affect the underlying structural properties of the material, while significantly improving its corrosion resistance and reducing hydrogen absorption during operation, without having to change the composition of the existing Zr alloys, a strategy that can be realized in a short period of time and at relatively low cost, and which represents one of the main directions of research at present.

In order to achieve the goals of safety performance metrics under extreme conditions, especially slowing down the oxidation rate kinetics of Zr alloy cladding materials in high-temperature vapor environments during severe accidents, Zr alloy coatings need to possess the following key properties:

(1) Low Neutron Absorption Cross-section: to ensure that the coating does not significantly affect the neutron yield of the nuclear reactor.
(2) Excellent Resistance to High-Temperature Vapor Oxidation: It can effectively prevent or slow down the oxidation of Zr alloys in high-temperature steam environments, providing more response buffer time under extreme accident conditions.
(3) Excellent Radiation Resistance: able to maintain the same performance under the radiation environment.
(4) Excellent Thermal Shock Resistance: able to maintain structural and performance stability under rapid temperature changes.
(5) Good Compatibility with the Zr Alloy Matrix: including matching of the thermal expansion coefficient, good thermal conductivity, and suitability of the coating thickness to prevent cracking and peeling of the coating during service [2].

For this reason, researchers look for materials that are resistant to high-temperature vapor oxidation. Figure 2.1 compares the parabolic oxidation rates of several common coating materials in a high-temperature vapor environment. Comparison results show that Cr_2O_3, Al_2O_3, and SiO_2 exhibit good stability in high-temperature steam environments. The parabolic oxidation rate of materials forming these oxide films is approximately two orders of magnitude lower than that of other materials, and coatings containing these elements provide a relatively good protective effect.

At present, the coating materials used on the surface of Zr alloys are mainly divided into two main categories: metal coatings and ceramic coatings, both of which exhibit outstanding antioxidant capabilities at high temperatures. Metal and alloy coatings mainly include Fe-based alloys, Cr, Cr-Al, Ti, and so on. These metal coatings usually provide superior mechanical properties and corrosion resistance,

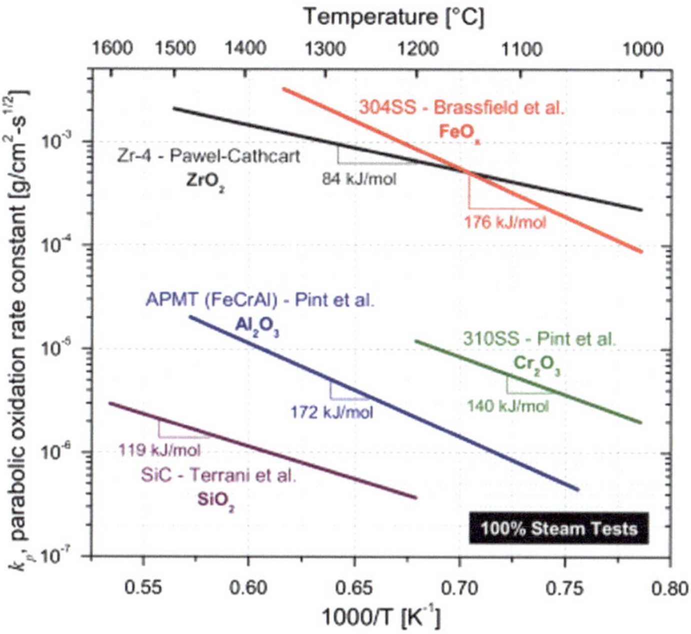

Fig. 2.1 Parabolic oxidation rates of relevant materials and oxidation kinetics curves in high-temperature steam environments [3]

especially oxidation resistance at high temperatures. Ceramic coatings are widely studied and applied because of their excellent corrosion and thermal shock resistance. They mainly include MAX-phase materials and nitrides such as Ti_3AlC_2, Ti_2AlC, Cr_2AlC, Zr_2AlC, and Zr_2SiC. The following are the current research status of several coating technologies with promising applications.

2.1.1 Cr Coating

Chromium (Cr) coatings, as candidates for ATF accident fuel coatings, have several advantages. Firstly, the melting point of Cr is as high as 1907 °C, which is able to maintain stability under the high-temperature environment during the operation of nuclear reactors. Cr and Zr alloys have similar coefficients of thermal expansion (coefficients of thermal expansion: Cr 6.2×10^{-6} K^{-1}, Zr 5.8×10^{-6} K^{-1}), which can help to reduce the internal stresses due to the change of temperature, and to avoid cracking and detachment of coatings from the substrate. In addition, Cr can form a stable and dense Cr_2O_3 oxide film at high temperatures, and this dense oxide film can effectively prevent the diffusion of oxygen into the Zr alloy substrate, significantly reduce the oxidation rate of Zr alloy, and improve its antioxidant performance in high-temperature vapor environment. At the same time, the Cr coating also has

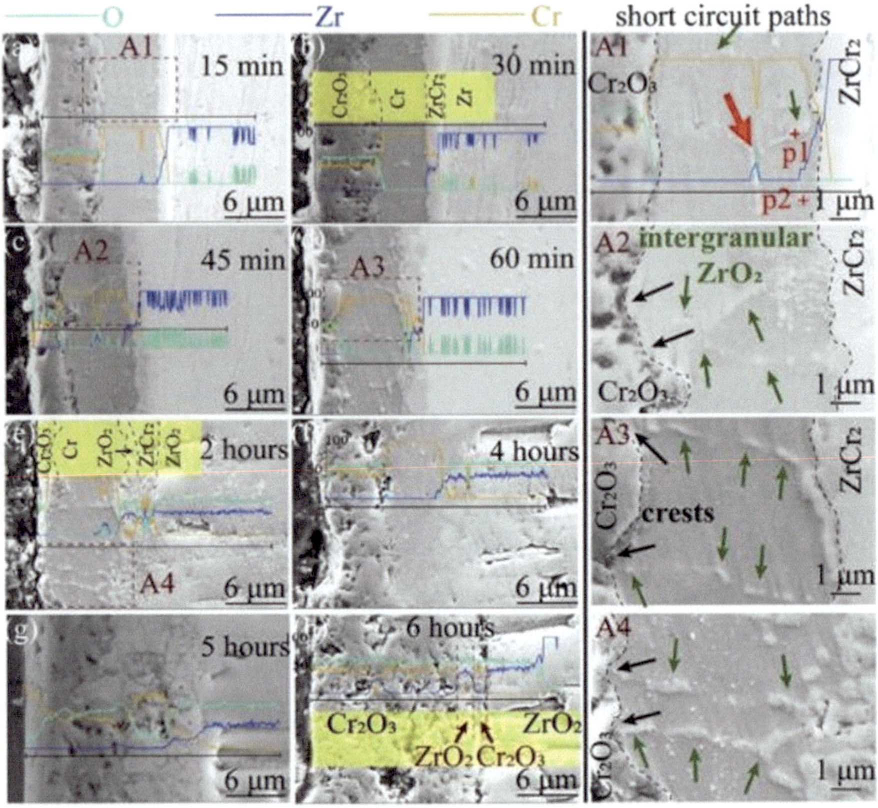

Fig. 2.2 SEM cross-sectional microstructure of Cr-coated Zr-4 alloy oxidized at 1200 °C for 15 min-6 h in a steam environment [5]

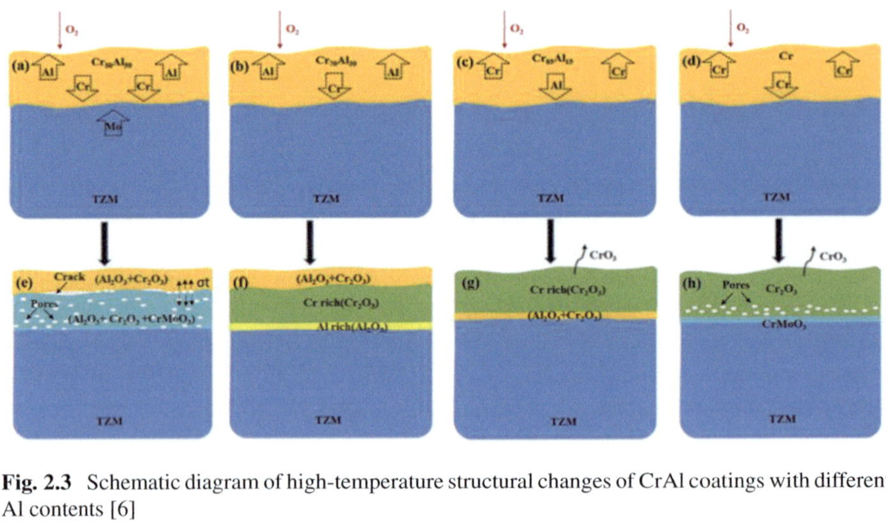

Fig. 2.3 Schematic diagram of high-temperature structural changes of CrAl coatings with different Al contents [6]

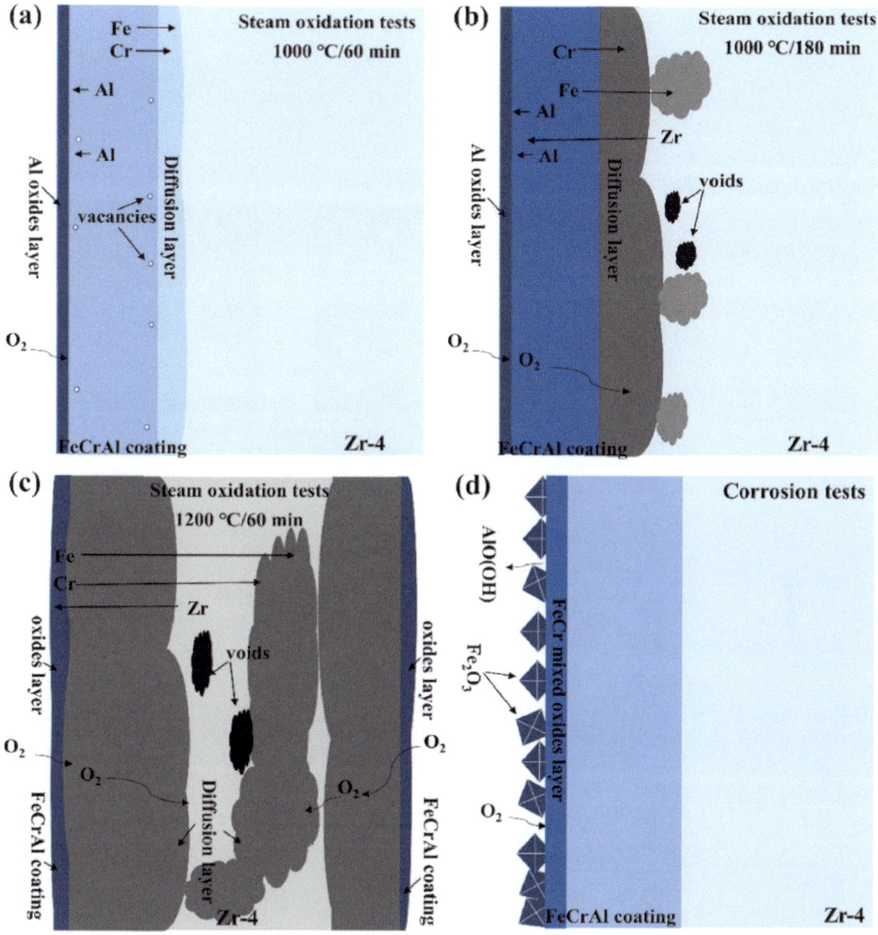

Fig. 2.4 Schematic representation of the microstructural evolution of FeCrAl coatings: after 60 min of vapor oxidation at 1000 °C (**a**), after 180 min of vapor oxidation at 1000 °C (**b**), after 60 min of vapor oxidation at 1200 °C (**c**), and after 50 days of autoclave hydrothermal corrosion testing at 360 °C and 18.7 MPa pressure (**d**) [7]

excellent corrosion resistance, which can protect the Zr alloy from corrosive media and enhance the overall durability of the fuel cladding [4].

Zhou et al. [5] prepared a 12-μm-thick Cr coating on Zr-4 alloy using a magnetron sputtering scheme and oxidized it for 1 h in a 1200 °C vapor environment, with no significant ZrO_2 generation observed. During the first 15–30 min, a four-layer structure including Cr_2O_3, residual Cr, interdiffusion layer, and β-Zr appeared in the surface layer of the coating and the Zr-4 alloy substrate. Meanwhile, due to the diffusion of elements, Kirkendall pores appeared in the coating. As the oxidation time prolonged, the residual Cr layer underwent re-oxidation. The newly formed Cr_2O_3

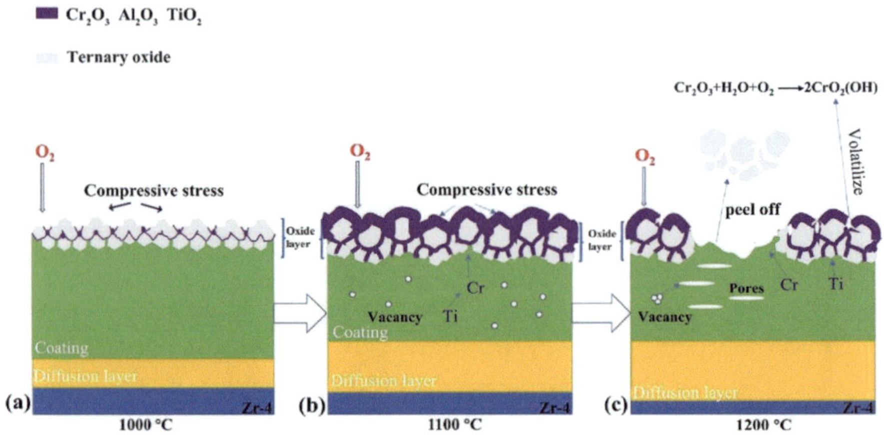

Fig. 2.5 Schematic diagram of oxidation process mechanism of AlCrNbTiZr coating

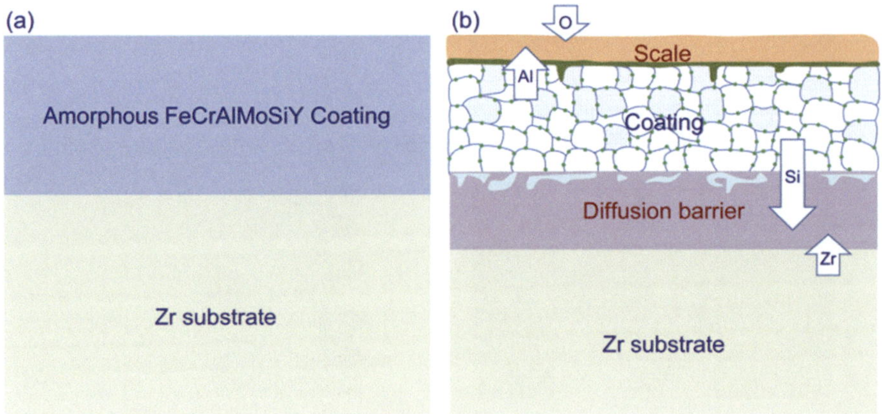

Fig. 2.6 Schematic diagram of microstructure evolution of coating vapor oxidation

exhibited a highly porous structure and lower density compared to the initially formed oxide layer. Notably, the residual Cr was fully oxidized during the 6-hour oxidation process, after which the oxidation rate of the Zr-4 alloy matrix became comparable to that of the uncoated Zr-4 alloy.

Although the Cr coating can effectively protect the Zr alloy substrate from oxidation at the initial stage, the protective efficacy is gradually weakened with the passage of time and the change of the oxidized layer. Therefore, in order to improve the long-term stability and safety of nuclear fuel cladding, the Cr coating needs to be improved. To this end, researchers have added other elements to optimize the protective effect of Cr coatings.

Qiu et al. [6] prepared 3–4-μm-thick CrAl coatings on TiZrMo alloys (TZM) by adding Al elements to the Cr coatings using a multi-arc ion plating method.

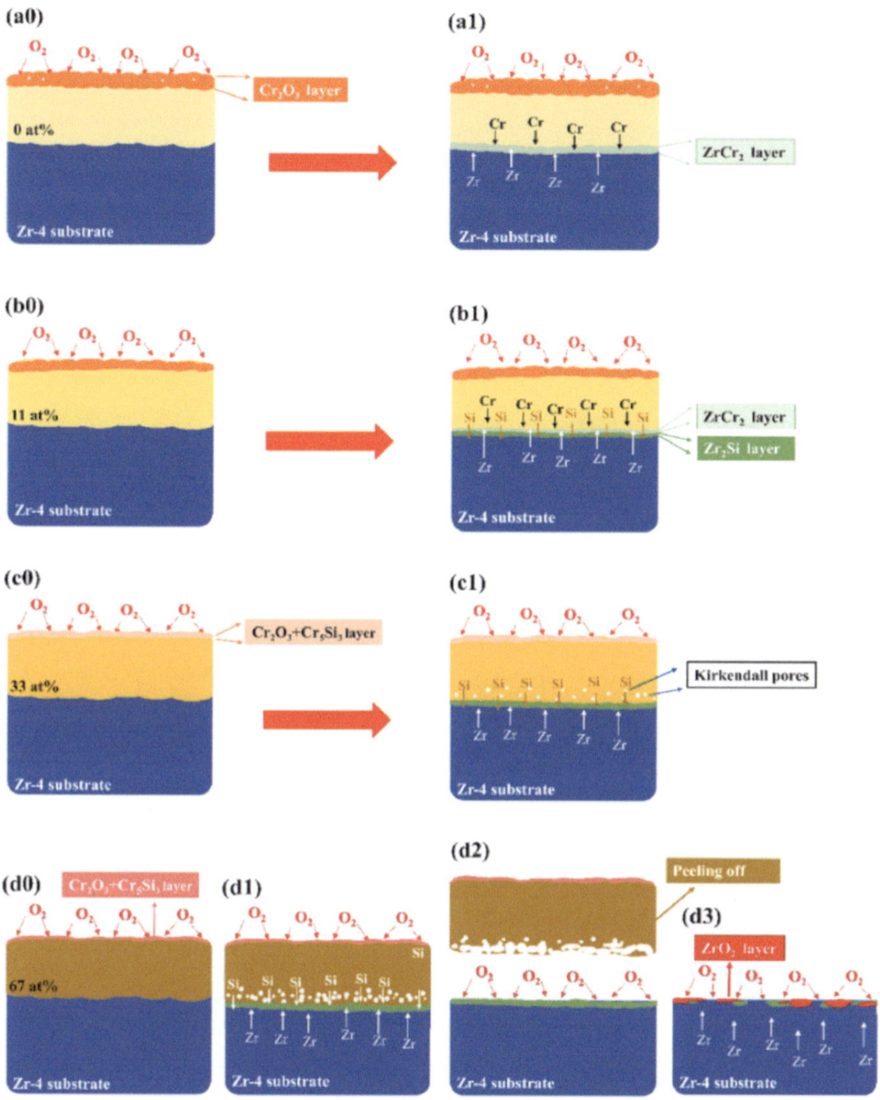

Fig. 2.7 Schematic of oxidation of CrSi-coated Zr-4 substrate during oxidation at 1100 °C: **a0–a1** 0 at%, **b0–b1** 11 at%, **c0–c1** 33 at%, **d0–d3** 67 at%

Under the high-temperature environment, the CrAl coating has higher hardness and excellent oxidation properties than the pure Cr coating due to the formation of a large number of Al-rich intermetallic compounds. The oxidation layer of the $Cr_{70}Al_{30}$ coating is divided into three parts, the surface mixed layer of Al_2O_3 and Cr_2O_3, the intermediate chromium-rich oxide (Cr_2O_3) layer, and the inner aluminum-rich oxide (Al_2O_3) layer, and the triple oxide layer enables it to exhibit better antioxidant

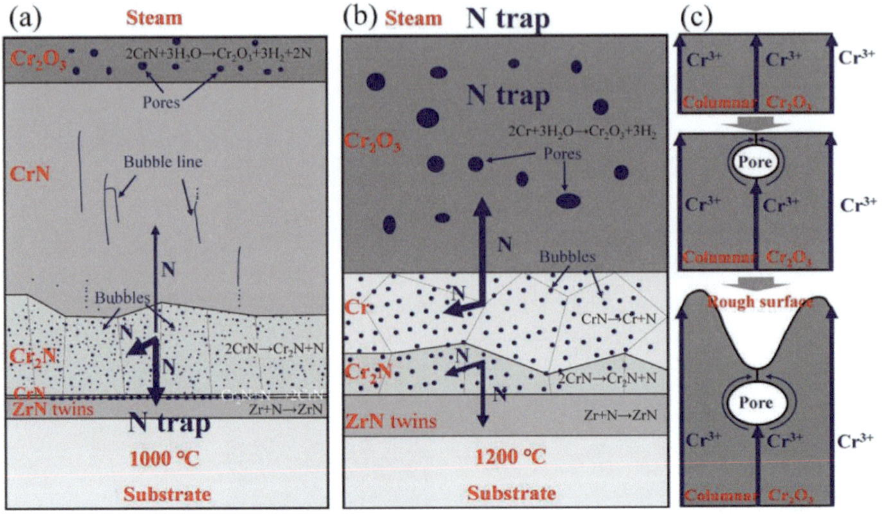

Fig. 2.8 Schematic representation of nitrogen diffusion within CrN coatings during vapor oxidation at 1000 and 1200 °C

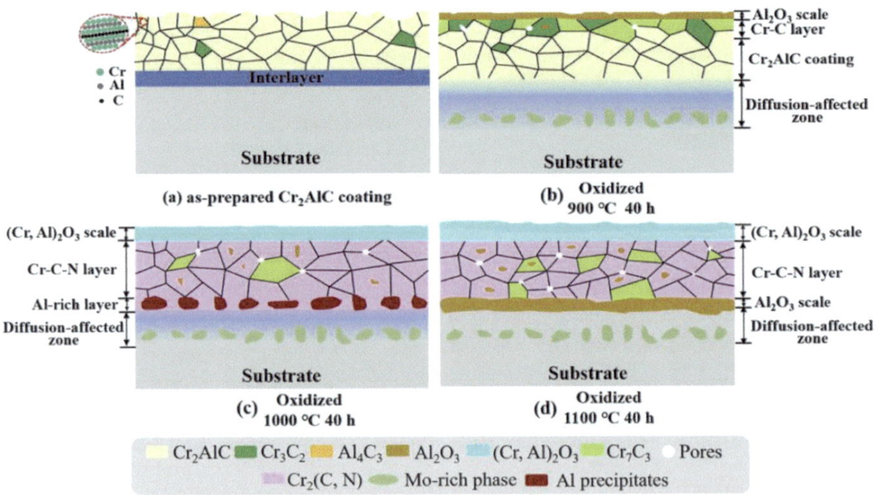

Fig. 2.9 Schematic representation of structural changes in Cr2AlC coatings at different temperatures

properties. Samples with an elemental Al content of less than 20 at.% are depleted of Al in the coating mainly in the form of internal oxidation, so that there is not enough Al in the outer oxide layer to form an Al_2O_3 oxide layer. $Cr_{50}Al_{50}$ coatings need to be supplied with Al from the inner coating due to the reaction between Al and O. The majority of the Al in the vicinity of the surface of the coating is depleted by

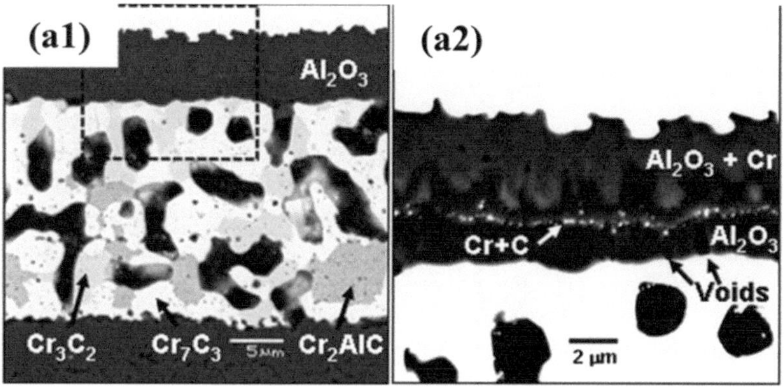

Fig. 2.10 Cross-section of Cr_2AlC film after oxidation at 1320 °C

the reaction between Al and O. Al_2O_3 oxide layer formation and Al diffusion are accompanied by a large number of microcracks and micropores. This makes it easier for O to diffuse through the coating into the substrate, leading to severe oxidation of the substrate. At the same time, a $CrMoO_3$ oxide layer is created between the outer oxide layer and the TZM substrate due to elemental diffusion at high temperatures. Many tiny pores appear at the interface between the oxide layer and the substrate. When the pure Cr coating is continuously oxidized in a high-temperature oxidative environment, the Cr_2O_3 layer reacts with oxygen to produce volatile CrO_3, and the volatilization of CrO_3 leads to the formation of micropores in the coating, which provide a channel for oxygen diffusion. The voids created due to Cr diffusion into the Mo matrix within the $CrMoO_3$ layer are conducive to the further diffusion of oxygen into the TZM alloy matrix.

Although the Cr coating effectively protects the Zr-alloy substrate from oxidation during the initial stage, its protective efficacy gradually diminishes with prolonged oxidation time and structural changes in the oxide layer. Therefore, to enhance the long-term stability and safety of nuclear fuel cladding, modifications to the Cr coating are necessary. To this end, researchers have incorporated additional elements to optimize the protective performance of Cr-based coatings.

2.1.2 *FeCrAl Coating*

FeCrAl coatings are potential candidates for accident-tolerant fuel protection coatings due to their high thermal conductance and excellent high-temperature oxidation resistance capabilities. The oxidation resistance capabilities of FeCrAl coatings at high temperatures mainly originate from the action of Al and Cr elements. At low temperatures, the top forms a continuous Cr_2O_3 with an internal Al_2O_3 layer; at high temperatures, Al is able to form a continuous and dense Al_2O_3 oxygen-barrier layer on the alloy surface, effectively preventing the residual coating from reacting with oxygen in contact, thus providing excellent antioxidant protection [7].

Liu et al. [8] used magnetron sputtering to deposit FeCrAl coating on the surface of Zr alloy, which was oxidized at 1000 °C in a vapor environment for 60 min, and a dense Al_2O_3 layer was formed on the surface, which had a better protective effect, but the diffusion of Fe and Cr elements in the coating to the Zr alloy substrate was serious, and the coating failed at longer time and higher temperature. Peng et al. [9] used plasma electrolytic oxidation to prepare a ZrO_2 buffer layer, and magnetron sputtering on ZrO_2 to prepare FeCrAl coatings, resulting in a two-layer ZrO_2/FeCrAl coating. The ZrO_2 buffer layer effectively mitigated the high-temperature interdiffusion between FeCrAl and Zr-4 alloys, which improved the adhesion and integrity of the FeCrAl layer after the vapor oxidation at 1000 °C.

Despite the excellent high-temperature antioxidation performance of the FeCrAl coating, its stability is limited by the service temperature. At high temperatures, Fe elements diffuse into the Zr alloy matrix and form Fe-Zr intermetallic compounds. These intermetallic compounds have large brittleness and poor ductility, which can lead to embrittling of the cladding and affect the overall performance and safety of the material. Meanwhile, when the oxidation temperature exceeds the Fe-Zr eutectic temperature, the FeCrAl coating will fail rapidly. This is because at high temperatures, the Al element in the coating diffuses into the base material, leading to a degradation of coating performance. At the same time, the high oxygen partial pressure of oxygen at the surface makes it easier for Cr_2O_3 to be oxidized into volatile CrO_3 at a faster rate and accumulate vacancies near the metal/oxide interface. At 1200 °C, the oxidative weight gain of FeCrAl coatings exceeds that of other alloys, and the coatings fail under extreme conditions [10, 11].

2.1.3 High-Entropy Coating

In 1995, Yeh et al. [12] proposed a new concept of alloys, high-entropy alloys (HEAs), which initially referred to alloys with single-phase structures composed of five or more metal elements in approximately equal atomic ratios. "Entropy" refers to the entropy of mixing (ΔS_{mix}), and researchers have taken ΔS_{mix}>1.5R as the limit for determining high entropy, believing that high entropy of mixing is conducive to inhibiting the generation of intermetallic compounds and thus stabilizing the solid solution structure. However, as the research process advances, the definition of HEAs is no longer shackled to criteria such as the number of group elements, iso-atomic ratio, and high conformational entropy. For the flexibility of HEAs design, researchers have relaxed the iso-atomic requirements, widened the allowable concentration range to 5–35 at.%, and found that some non-iso-ratio multi-major element alloys also exhibit excellent properties. Due to the high-entropy effect, lattice distortion effect, slow diffusion and cocktail effect, HEAs have properties unmatched by conventional alloys, including high hardness, wear resistance, corrosion resistance, and thermal stability, and therefore have considerable potential for application in surface protection.

When the service temperature of a metal material exceeds 0.6 times its melting point, the strength of the alloy decreases sharply. Traditional alloys exhibit suboptimal performance in high-temperature extreme environments such as aerospace and nuclear reactors. Therefore, the design of new alloys necessitates the preferential selection of metal elements with high melting points to adapt to these extreme conditions. Based on HEAs Senkov et al. [13] in the HEAs on the basis of refractory high-entropy alloys (Refractory High-entropy Alloys) (The concept of Refractory High-entropy Alloys (RHEAs) was introduced based on HEAs. In contrast to the traditional HEAs, where the elemental composition is chosen in the region of transition metals, RHEAs shift the perspective to metals with high melting points. People usually refer to the elements from the fourth to the sixth period of the periodic table and the fourth to the sixth subgroup as refractory metals, and RHEAs are new materials that use refractory elements with high melting points, such as Ti, V, Cr, Zr, Nb, Mo, Hf, Ta, and W, as the main alloy elements, and at the same time, add doped non-refractory elements, such as C, Al, Si, and Ni, to improve the performance of their alloys. In recent years, researchers have tested a number of properties of RHEAs, such as mechanical properties, irradiation resistance, corrosion resistance, and oxidation resistance, and enhanced the material properties through compositional adjustments.

Xin et al. [14] prepared AlNbTiZr MEA coatings with different Al contents on N36 substrate using magnetron co-sputtering technique. The effect of Al content on the microstructure and properties of MEA coatings was investigated. The hardness and elastic modulus of the coatings decreased and then increased with the increase of Al content, and the crack resistance and plastic deformation resistance were slightly improved. Compared with the uncoated N36 substrate, the AlNbTiZr MEA coating has a thinner oxide layer and stronger hydrophobicity for excellent corrosion resistance under pressurized water reactor operating conditions.

Yi et al. [15] prepared near-equimolar CrMoNbZr coatings with low thermal neutron absorption cross-section on zirconium alloy-4 substrate using RF magnetron sputtering at four different temperatures, and results showed that the corrosion resistance of the coating mechanism deposited at 300 °C is superior to that of the coatings prepared at other temperatures.

He et al. [16] prepared AlCrNbTiZr coatings on Zr-4 substrates at different deposition temperatures and showed that the coatings prepared under RT conditions have better high-temperature corrosion resistance in pressurized water reactor environments, and the RT-deposited coatings have good oxidation resistance after the high-temperature oxidation tests at 1000 °C-1200 °C. The co-diffusion effect of Cr and Ti is the main reason for the oxidation of the coating. The co-diffusion effect of Cr and Ti is the main reason for the decrease in the oxidation performance of the coating. In addition, the interdiffusion between the coating and the Zr-4 substrate is the reason for the enhanced substrate protection.

Li et al. [17] sputtered an amorphous FeCrAlMoSiY coating on the surface of a zirconium alloy, and during the vapor oxidation process at 1200 °C, Si atoms preferentially migrated to the coating-substrate interface to form a continuous Zr_2-rich Si diffusion-barrier layer to inhibit inward diffusion of the coating and outward diffusion of the substrate elements. As a result, the formation of α-Al_2O_3 protective layer was promoted, effectively preventing the subjacent coating and substrate from being oxidized.

Zhao et al. [18] prepared a near-equimolar ratio high-entropy alloy (HEA) AlTiCrNiTa coating on Zr-4 substrate using radio frequency (RF) magnetron sputtering (MS) technique, and the results show that the structure of AlTiCrNiTa HEA coating is basically amorphous with the presence of FCC nanocrystalline domains, and after 45 days of high-pressure sterilization test, the AlTiCrNiTa HEA coating still effectively protects the Zr-4 substrate from oxides.

Liu et al. [19] prepared (TiVZrCrAl)N high-entropy nitride coatings on Zr-4 alloy substrates by reactive magnetron sputtering technique. The results show that the coating thickness decreases and then increases with the decrease of the N_2/Ar pressure ratio, and the hardness increases and then decreases; all the coatings show B_1-NaCl structure. The (TiVZrCrAl)N coatings have excellent resistance to high-temperature corrosion and vapor oxidation when $I = 1/12$.

The outward diffusion of alloying elements may lead to the formation of a large number of Kirkendall pores in the substrate, making the coating structurally undense during oxidation, and then endangering its mechanical properties. Also, for most RHEAs, thickened oxides tend to form on the surface, but many metal oxides melt or evaporate at specific temperatures and are not protective, which affects the structure of the oxide layer. In addition, the complexity of the preparation process of high-entropy alloys and the high price of refractory metals as well as transition metal elements make the preparation cost high, limiting industrial applications.

2.1.4 Ceramic Coating

Ceramic coatings are mainly SiC, nitride, and other directions. The protective mechanism is that the Al, Cr, Si, and other elements in the coating generate a dense oxide barrier layer on the surface at high-temperature environments, thus providing protection in high-temperature environments.

Metal silicides are well-established in mechanical applications: for example, $MoSi_2$ is utilized in heating elements, flame stabilizers, igniters, and hard coatings, while $NbSi_2$ finds use in thermal-barrier coatings and environmental-barrier coatings. $NbSi_2$ is used in thermal-barrier coatings and environmental-barrier coatings. In addition, silicides are also used in thermoelectric devices ($CrSi_2$), field emitters ($TiSi_2$), refractory materials ($ZrSi_2$), and so on.

Yang et al. [20] prepared CrSi coatings using magnetron sputtering method. The coatings were harder than the Cr coatings, but the bonding with the substrate was a little worse. Samples with 11 At.% Si content were oxidized at high temperatures, the coatings produced a Zr_2Si diffusion layer with the substrate, which mitigated the Cr-Zr diffusion, the surface of the coatings produced a dense Cr_2O_3 oxygen-barrier layer, and the internal Cr_3Si phase of the coatings prevented the diffusion of O and inhibited the reaction between $ZrCr_2$ and O. The coating was then oxidized at 1100 °C, which resulted in the formation of the $ZrSi_2$ coating, and the $ZrCr_2$ coating was then oxidized to a higher temperature. This makes the coating oxidized for half an hour at 1100 °C still have protective effect.

Silicide coatings have excellent resistance to high-temperature oxidation, however, in tests simulating the operating environment of the cladding (a high temperature of 360 °C and high-pressure water environment), leaching of Si occurs, resulting in reduced or ineffective coating protection, as well as brittleness and low-temperature oxidation of the coatings, which limits their various other applications.

In nitride coatings, the Al element is introduced into the TiN and CrN coatings, which have high hardness, melting point, and thermal conductance, and a continuous dense Al_2O_3 oxygen-barrier layer is formed on the surface in high-temperature environments to improve oxidation resistance.

Yun et al. [21] prepared CrN coatings using cathodic arc ion plating method. A thin ZrN nanotwin layer was formed at the interface between the coating and the substrate, which improved the adhesion between the coating and the substrate, and the coating had the antioxidant property of protection for 1 h in 1200 °C vapor environment. However, the decomposition of CrN and oxidation of nitrides under high-temperature vapor environment released nitrogen atoms, which formed nitrogen bubbles inside the coating, and these bubbles negatively affected the structural integrity of the coating.

Liu et al. [22] prepared CrAlN coatings using a multi-arc ion plating method. The coating formed a Cr_2O_3 and Al_2O_3 oxidation-resistant barrier layer on the surface under the vapor environment at 1000 °C, and it still had a protective performance after 2 h of oxidation. However, when the temperature rises to 1200 °C, the nitrogen element in the coating volatilizes, leaving nitrogen bubbles in the coating, and these bubbles become crack generating points, the cracks and thermal stresses lead to brittle fracture of the coating, and the coating protection fails.

Although the nitride coating has the advantages of good oxidation resistance in high-temperature environment, the coating in the high-temperature nitrogen gas leaves the pores, resulting in cracks in the coating. Additionally, in a high-temperature, high-pressure water environment at 360°C, the adhesion of oxidized corrosion products is relatively weak, and they consist of soluble magnetite (AlOOH) material. This would lead to cracking and peeling of the coating, resulting in failures [23].

2.1.5 MAX-Phase Coating

The MAX phase represents a class of layered ternary compounds with the general formula Mn+1AXn, where M is a transition metal, A is a third or fourth-group element, X is carbon or nitrogen, and n ranges from 1 to 3. Ti_2AlC and Cr_2AlC are typical MAX-phase materials, and the mechanism of protection in high-temperature environments is the formation of a continuous and dense oxygen-barrier layer of oxides by the Cr and Al elements in the coating.

Ke et al. [24] prepared Cr_2AlC coatings using a combined arc/sputtering technique. The selective oxidation of Al elements in high-temperature environment generated Cr_2O_3 and Al_2O_3 oxygen-barrier layer on the surface, and In high-temperature environments, aluminum undergoes selective oxidation, forming an oxide layer on the surface as protection and internal phases such as Cr_7C_3 and Cr_3C_2 phases. At the same time, Al elements are enriched on the surface of the coating as well as at the coating/substrate interface, and when the oxidation temperature rises to 1100 °C, the Al-rich interfacial layer transforms into a dense α-Al_2O_3 layer. These structures enable the coatings to provide good protection at 1100 °C.

Hajas et al. [25] investigated the isothermal oxidation behavior of Cr_2AlC coatings on alumina substrates in the temperature range of 1230–1410 °C. The consumption of Al and oxidation of Cr carbides resulted in the formation of CO gas and Cr_2O_3. The simultaneous occurrence of gas formation and the depletion of Al and Cr led to the formation of a large number of voids observed in the chromium carbide layer.

Xu et al. [26] investigated the oxidation resistance of woven Ti_3AlC_2 ceramics in the temperature range of 1000–1300 °C. During the oxidation process, the woven top surface (TTS) formed at a temperature of 1473 K, which is parallel to the basal plane and orthogonal to the c-axis, was covered with a surface oxide consisting of a porous mixture of TiO_2 and Al_2TiO_5 (Fig. 2.11c_1–c_2). Due to the preferential outward diffusion of aluminum atoms on the (000 1) substrate, a dense passivated aluminum oxide layer is rapidly formed. As a result, a distinctly different microstructure was observed on the TSS with a surface consisting of dense aluminum oxide (Fig. 2.11c_3–c_4). The results show that there is a significant anisotropy in the oxidation process, and the samples exhibit good oxidation resistance when tested along the c-axis.

Tang et al. [27] investigated the oxidation behavior of bulk Ti_2AlC ceramics in steam at temperatures from 1400 to 1600 °C. With the increase of temperature, delamination of the oxide layer occurs and a large amount of volatile gases gradually transforms the dense oxide layer into a loose porous structure. During steady-state oxidation, Ti and C diffuse outward through the α-Al_2O_3 grain boundaries, leading to the formation of gaseous CO and CO_2 by the deviation of TiO_2 at the α-Al_2O_3 grain boundaries. The flux rates of H_2 and CO_2 in the exhaust gases were significantly increased as compared to the oxidation at 1400 °C. The oxidation of the MAX-phase coatings in steam was also observed at the same temperature as that of the metal coatings.

The MAX-phase coatings have the advantages of high thermal conductivity of metal coatings and stability in neutron irradiation environments, though. However, problems such as weak oxidation resistance in high-temperature environments (above 1000 °C) need to be improved.

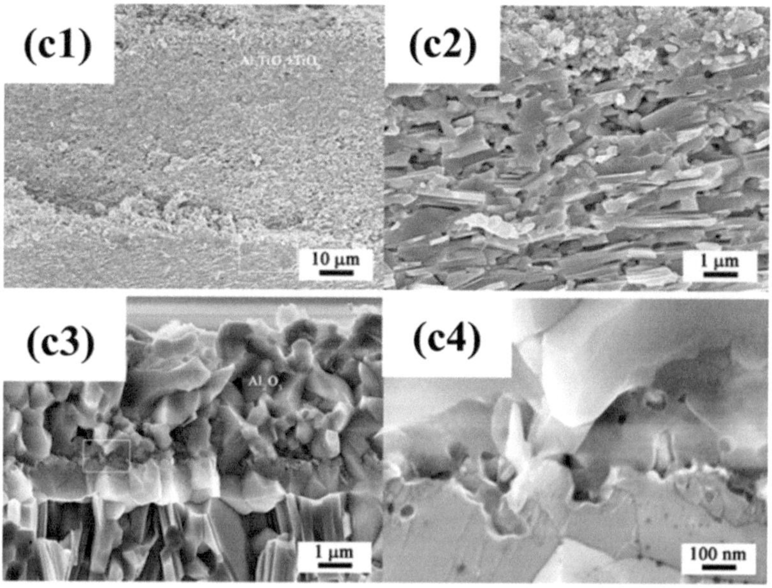

Fig. 2.11 Fracture surfaces of scales on organized Ti_3AlC_2 ceramics

2.2 Conclusion

This chapter discusses the development of protective coatings for zirconium alloy fuel cladding, which are crucial for enhancing the safety and performance of nuclear reactors under extreme conditions, such as Loss-of-Coolant Accidents (LOCA), and aim to improve the oxidation resistance of accident-tolerant fuel (ATF) technology. The main research strategies include the introduction of new antioxidant materials and surface modifications of existing Zr alloys. Chromium coatings can form a dense layer of chromium oxide, but long-term stability and volatility are issues. FeCrAl coatings can form an alumina layer, but the interdiffusion with Zr requires a ZrO_2 buffer layer to address the problem. High-entropy alloy coatings have excellent performance but are complex to prepare and costly. Ceramic and MAX-phase coatings provide good oxidation protection but have limitations in high-pressure water and high temperatures. Despite the potential of these coatings, issues of stability, diffusion, and cost need to be resolved for their application in advanced nuclear reactors.

References

1. Kim HG, Kim IH, Jung YI et al (2018) Out-of-pile performance of surface-modified Zr cladding for accident tolerant fuel in LWRs. J Nucl Mater 510:93–99
2. Younker I, Fratoni M (2016) Neutronic evaluation of coating and cladding materials for accident tolerant fuels. Prog Nucl Energy 88:10–18
3. Terrani KA (2018) Accident tolerant fuel cladding development: Promise, status, and challenges. J Nucl Mater 501:13–30
4. Han X, Chen C, Tan Y et al (2020) A systematic study of the oxidation behavior of Cr coatings on Zry4 substrates in high temperature steam environment. Corros Sci 174:108826
5. Meng Y, Chen C, Zeng S et al (2024) Investigations of oxidation behavior and establishment of life-cycle model during the steam oxidation of Cr-coated Zry-4 at 1200 °C. Corros Sci 226:111694
6. Wang H, Ren H, Qiu L et al (2023) Effect of Al content on 1200 °C steam oxidation behavior of Cr-based coatings on TZM alloy. Corros Sci 219:111239
7. Chikhalikar A, Roy I, Abouelella H et al (2022) Effect of aluminum on the FeCr (Al) alloy oxidation resistance in steam environment at low temperature (400 °C) and high temperature (1200 °C). Corros Sci 209:110765
8. He L, Liu C, Lin J et al (2021) Microstructure, oxidation and corrosion properties of FeCrAl coatings with low Al content prepared by magnetron sputtering for accident tolerant fuel cladding. J Nucl Mater 551:152966
9. Wang Y, Tang H, Han X et al (2018) Oxidation resistance improvement of Zr-4 alloy in 1000 °C steam environment using ZrO_2/FeCrAl bilayer coating. Surf Coat Technol 349:807–815
10. Dabney T, Johnson G, Yeom H et al (2019) Experimental evaluation of cold spray FeCrAl alloys coated zirconium-alloy for potential accident tolerant fuel cladding. Nuclear Materials and Energy 21:100715
11. Gigax JG, Kennas M, Kim H et al (2019) Interface reactions and mechanical properties of FeCrAl-coated Zircaloy-4. J Nucl Mater 519:57–63
12. Yeh JW, Chen SK, Lin SJ et al (2004) Nanostructured high-entropy alloys with multiple principal elements: novel alloy design concepts and outcomes. Adv Eng Mater 6(5):299–303
13. Senkov ON, Wilks GB, Miracle DB et al (2010) Refractory high-entropy alloys. Intermetallics 18(9):1758–1765
14. Xin H, Yang J, Mao J et al (2021) Effect of Al concentration on microstructure and properties of AlNbTiZr medium-entropy alloy coatings. Materials 14:7661
15. Yi J, Zhao J, Liu C et al (2023) Effect of deposition temperature on the hightemperature corrosion resistance of CrMoNbZr medium-entropy alloys coatings. Adv Eng Mater 25:2300051
16. He H, Lai M, Liu C et al (2023) Effects of deposition temperature on the microstructure, mechanical properties, high temperature corrosion and oxidation properties of AlCrNbTiZr high entropy alloy coatings. Int J Refract Metals Hard Mater 106156
17. Li Y, Meng F, Ge F et al (2021) Improved oxidation resistance through an in-situ formed diffusion barrier: Oxidation behavior of amorphous multi-component FeCrAlMoSiY-coated Zr in high-temperature steam. Corrosion Sci 109566
18. Zhao S, Liu C, Yang J et al (2021) Mechanical and high-temperature corrosion properties of AlTiCrNiTa high entropy alloy coating prepared by magnetron sputtering for accident-tolerant fuel cladding. Surface and Coat Technol 127228
19. Liu S, Liu C, Yang Z et al (2024) Microstructure, high-temperature corrosion resistance and oxidation properties of (TiVZrCrAl)N high entropy nitride coatings with different N_2/Ar ratios. Surf Coat Technol 130226
20. Zhao S, Zhong Y, Li Z et al (2023) Effect of Si content on the microstructure and properties of CrSi coatings for ATF cladding. Vacuum 218:112627
21. Liu J, Hao Z, Cui Z et al (2021) Oxidation behavior, thermal stability, and the coating/substrate interface evolution of CrN-coated Zircaloy under high-temperature steam. Corros Sci 185:109416

22. Liu J, Cui Z, Ma D et al (2020) Investigation of oxidation behaviors of coated Zircaloy as accident-tolerant fuel with CrAlN and CrAlSiN coatings in high-temperature steam. Corros Sci 175:108896
23. Alat E, Motta AT, Comstock RJ et al (2015) Ceramic coating for corrosion (c_3) resistance of nuclear fuel cladding. Surf Coat Technol 281:133–143
24. Wang Z, Ma G, Liu L et al (2020) High-performance Cr_2AlC MAX phase coatings: oxidation mechanisms in the 900–1100 °C temperature range. Corros Sci 167:108492
25. Hajas DE, to Baben M, Hallstedt B et al (2011) Oxidation of Cr_2AlC coatings in the temperature range of 1230 to 1410 °C. Surf Coat Technol 206:591–598
26. Xu L, Zhu D, Liu Y et al (2018) Effect of texture on oxidation resistance of Ti_3AlC_2. J Eur Ceram Soc 38:3417–3423
27. Tang C, Steinbrück M, Große M et al (2017) Oxidation behavior of Ti_2AlC in the temperature range of 1400–1600 °C in steam. J Nucl Mater 490:130–142

Chapter 3
Advantages and Bottlenecks of Conventional Coating Fabrication Technology

Abstract In order for zirconium cladding coatings to provide effective protection under accidental working conditions, the coating surface must be uniform, dense, and have good resistance to high-temperature oxidation. It is particularly important to control the coating properties through the use of different deposition techniques. This chapter discusses the application of various coating technologies such as cold spraying, magnetron sputtering, arc ion plating, and laser surface modification in the preparation of coatings. It highlights the advantages and disadvantages of these technologies, as well as directions for future research. Compared with other coating preparation techniques, magnetron sputtering has been widely used in the preparation of zirconium alloy cladding coatings due to its wide range of target material selection, small workpiece deformation, controllable deposition rate, good process repeatability, and many other advantages, and has been widely used in industrial applications.

Keywords Coating preparation · Spraying · Electric arc ion plating · Magnetron sputtering · Laser cladding

3.1 Spraying

Spraying mainly includes cold spraying, thermal spraying, and plasma spraying technology; thermal spraying is a surface coating technology, including flame spraying, arc spraying, etc., by heating the material to the melting point or semi-molten state, and then sprayed on the surface of the work piece to form a coating, in order to improve the surface properties of the material. Thermal spraying method is different from other coating deposition methods, does not use individual ions, atoms, molecules form the deposition of coatings, but through the heating of the dispersion of finely divided coating material to the melting or semi-molten state, with high-speed gas carrying and blowing the heated material impact on the surface of the pre-treated substrate, the formation of coatings. The coating material is mostly in the form of powder, wire, or rod. Liu et al. [1] prepared WC/Ni60 coatings on 316L stainless

J. Hu and G. Cao, *Zirconium Alloy Coatings*,
https://doi.org/10.1007/978-981-96-5028-6_3

steel to improve the wear resistance of the substrate in a variety of environments, the results show that the coating is a layered structure, mainly composed of γ-Ni solid solution, carbides and borides, and wear experiments have found that the surface of the coating formed a passivation film consisting of WC, $Ni(OH)_2$, and FeOOH, which effectively improves the corrosion resistance of the coating. There are some problems with thermal spray technology. Duan et al. [2] prepared aluminum oxide (Al_2O_3) and zirconium oxide (ZrO_2) coatings with a thickness of 100 μm on the surface of Zr-2 alloys using thermal spraying technique and carried out homogeneous corrosion in a distilled water environment at 19.2 MPa and 360 °C, as well as electrochemical tests in 3.5 wt. % NaCl solution to evaluate the corrosion behavior of the coatings, and the results showed that both coatings were unable to prevent the electrolyte diffusion into the substrate and could not protect the substrate from corrosion. Thermal spraying is not suitable as a surface modification technique for zirconium alloys because of the high working temperature, the severe oxidation of the material during spraying, and the poor impact resistance due to the low bond strength of the coating to the substrate, which is mainly based on mechanical bonding.

Plasma spraying surface processing technology as a precision coating method is in the flame spraying after the gradual development. The heat source used in plasma spraying is a high-energy plasma, and the specific process of preparing the film layer is as follows: firstly, the heat source heats up the film material to a molten or semi-molten state, then, the flame stream is accelerated to a high speed, and finally, the film material is sprayed onto the surface of the substrate material, and then the film layer is obtained as required. This spraying method has the advantages of high deposition rate, easy preparation of thick layers, and wide range of available materials and has been used in the preparation of zirconium alloy surface coatings. For example, Wang et al. [3] used atmospheric plasma spraying to deposit a Cr layer with a thickness of about 70 um on a Zr-4 substrate, which was subjected to a 1200 °C /1h high-temperature water vapor oxidation experiment, and the results showed that the Cr layer prepared by plasma spraying improved the resistance to high-temperature oxidation of the zirconium substrate. Duan et al. [3] used plasma spraying to prepare Al_2O_3 and ZrO_2 coatings on the surface of Zr-2 substrate, in which the thickness of the film layer is about 100 μm, studied the changes in the hardness and corrosion resistance of the material before and after the preparation of the coating, and explored the mechanism of the expansion of cracks in the film layer. Plasma spraying prepared by the membrane layer there will be uneven distribution of pores, the coating preparation of the stress can not be fully released will lead to the formation of microcracks on the surface, the existence of large porosity, loose and other defects (Fig. 3.1) is the key factor in the failure of the membrane layer as well as in the spraying process of the dust and noise and other issues are to a certain extent to limit the thermal spraying technology in the zirconium alloy cladding coating applications (Fig. 3.2).

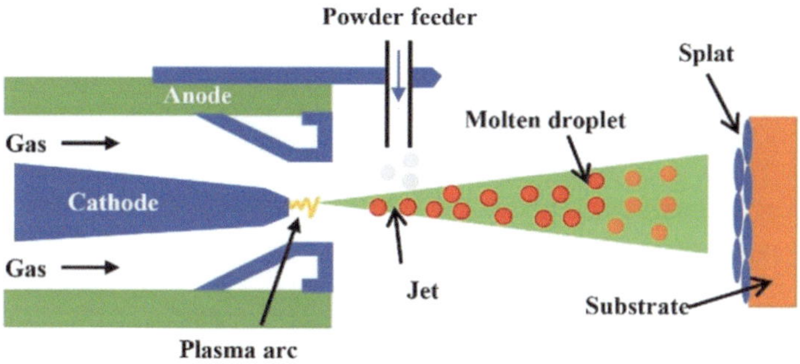

Fig. 3.1 Schematic diagram of the working principle of plasma spraying

Cold spraying, known as supersonic cold spraying, is a new surface coating technology for materials first introduced in the 1990s [6]. The principle is shown in Fig. 3.3, at room temperature or lower temperature, by supersonic gas, solid two-phase airflow will be coated powder shot to the substrate, when impact, the particles undergo plastic deformation damage and decomposition of the powder and the substrate surface of the oxide layer, resulting in metallurgical bonding and mechanical interlocking. The advantages of cold spraying are: low working temperature of the film layer preparation, there is no metal powder melting phenomenon; layer of particles spraying speed, with a high deposition rate; film layer is more dense, the film layer and the substrate combined with a good and so on.

Currently, cold spraying technology is used in aerospace, electronics and electric power, medical devices, energy power, weaponry, and other fields. Such as Lee [7] Cr coatings were prepared on Zr-4 substrates by cold spraying and physical vapor deposition techniques, respectively, and flow-boiling heat transfer tests were carried out, which showed that the Cr coatings deposited by cold spraying significantly improved the oxidation resistance of the substrates. Dong et al. [8] FeCrAl coatings were prepared on zirconium alloy plates using a cold spraying process with a film thickness of about 140 μm, and the samples were subjected to oxidation experiments at 1200 °C/3000 s to confirm the oxidation protection effect of the coatings on the zirconium alloy substrate, as well as to improve the ductility and mechanical strength of the zirconium alloy substrate. Maier et al. [9] Ti_2AlC coatings with a thickness of about 90 μm were prepared by cold spraying on the surface of Zr-4 plates. The cold spraying technique has a high deposition rate and is safe and green, but there are problems such as poor spraying direction, loose structure of the deposited coating, nozzle clogging, and surface unevenness. The future research directions include: (1) exploring the interaction between the cold sprayed material and the substrate; (2) trying to optimize the deposition rate by changing the structure of the powder and the oxidized state of the surface; (3) investigating the principle of the spraying from the thermodynamic aspect; (4) to study the organizational evolution of particles during preparation, etc.

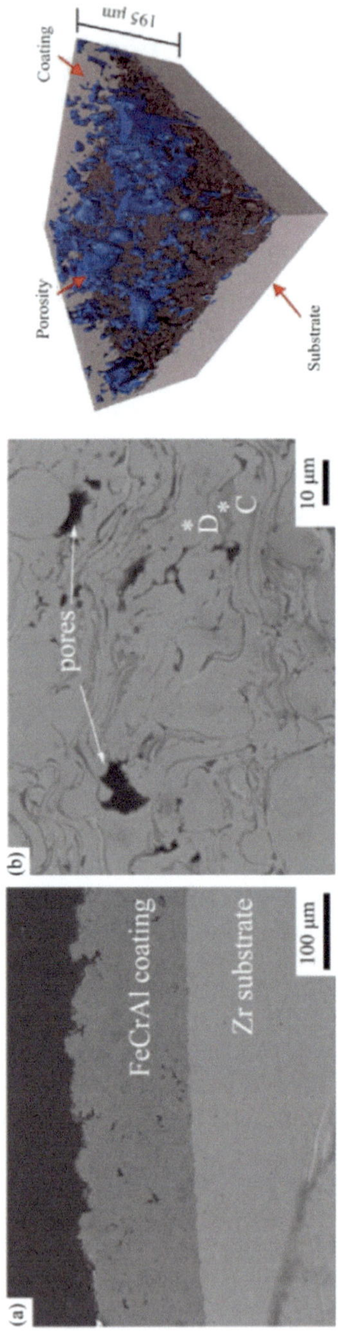

Fig. 3.2 Plasma sprayed FeCrAl film layers (**a**), (**b**) [4] and aperture penetration through the sprayed thick film in 3D stereo [5]

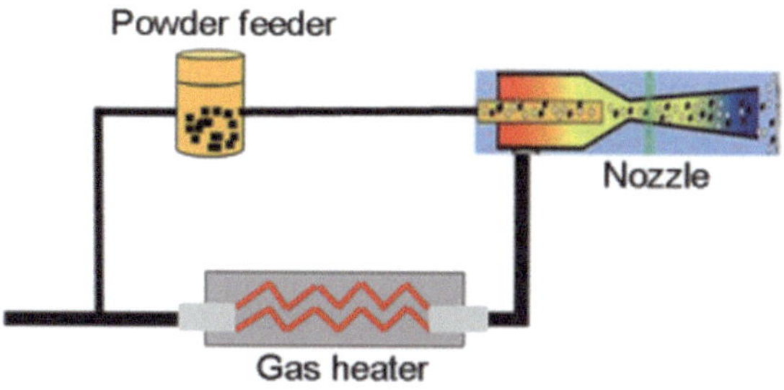

Fig. 3.3 Schematic diagram of the working principle of cold spraying

Color online

3.2 Electric Arc Ion Plating

Considering that the coating preparation process should avoid as much as possible the influence on the organizational structure of the zirconium matrix, the deposition temperature should be as low as possible. Therefore, the zirconium alloy surface coating preparation process can adopt the physical vapor phase deposition method with relatively low deposition temperature. Physical vapor deposition technology can better avoid the temperature rise of the substrate in the process of film deposition, which can be used for the preparation of zirconium alloy surface modification coating; the current use of more is the arc ion plating and magnetron sputtering technology.

Arc ion plating technology is used in a vacuum environment between the cathode (target) and the anode formed by the vacuum chamber to trigger an arc discharge, the arc movement evaporates the target and deposited on the substrate surface and. Arc generation is based on the phenomenon of high-speed changes in high-density current in a very small space. Arc ion plating technology is based on the principle of cold cathode vacuum discharge theory, which belongs to the physical vapor deposition (PVD) of a kind of PVD technology to prepare a dense film layer, and excellent performance can be effectively enhance the performance of the film layer of the traditional technology.

The arc (diameter 10^{-8}–10^{-4}m) around the surface of the metal target rapid movement (10^2m/s order of magnitude), current density up to (10^8–10^{12}A/m), so that the material to be coated evaporated. The evaporated material is almost completely ionized and forms a high-energy plasma. The metal ions combine with reaction gases (e.g., nitrogen N_2, oxygen O_2, acetylene C_2H_2, argon Ar, etc.) added to the reaction chamber to impact the tool or component to be plated with high energy and are finally deposited as a film layer. As shown in Fig. 3.4, high-energy plasma is injected, as well as solid and molten particles, and the ejected material exists as multivalent ions, which maintains evaporation and combustion.

Technical advantages of arc ion plating: ① plasma energy, high film adhesion; ② high ionization rate (60–90%), fast growth rate; ③ dense film, low porosity, etc.; ④ no auxiliary means of ionization, easy to control; ⑤ evaporation source is a solid cathode target, the arc target source can be set in any orientation, multisource arrangement to ensure uniformity of the film; ⑥ strong plating capacity around the plating, a

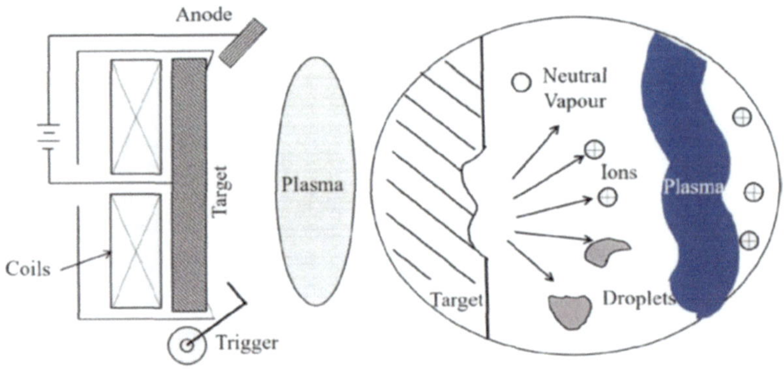

Fig. 3.4 Sketch of the operating principle of the cathodic arc

wide range of materials can be plated and can be deposited compound film. Because AIP has a variety of advantages, the market share in the field of tool film is as high as 60–70%, which fully demonstrates the maturity of the technology. At present, ion plating mainly includes magnetron sputtering ion plating, reactive ion plating, hollow cathode evaporation method, and multi-arc ion plating. Ion plating can be used to spray ceramic materials, metal materials, and their mixtures or compounds. For example Park et al. [10] Cr coatings were prepared on Zr-4 substrates by ion plating, and the Cr coatings exhibited excellent oxidation resistance in a high-temperature vapor environment at 1473 K. The Cr coatings were also deposited on Zr-4 substrates by ion plating.

Multi-arc ion plating is a method to further increase the rate of film deposition, thus using two or more arc evaporation sources to prepare the film layer, which is widely used in the work of zirconium alloy cladding surface coating preparation, and its main features include: high film strength, fast deposition rate, high bonding between the film layer and the substrate, high target dissociation rate, and better shot-winding properties Pl et al. [11] deposited CrN coatings by multi-arc ion plating and conducted oxidation experiments at 1160 °C on the film layers. The results show that the prepared film layer is uniform and dense, the average film thickness is about 13 μm, the film layer did not crack or flake off after oxidation, and the CrN coating shows excellent oxidation resistance and mechanical properties.

Arc ion plating is prone to large particle deposition in the sputtering process, uneven coating sputtering and other problems. Sun et al. [12] prepared a Cr coating on Zr-4 alloy by arc ion plating. A small amount of pores and small cracks were observed at the interface between the Cr coating and Zr-4 substrate by TEM, so it is still necessary to optimize the coating preparation process, for example, using magnetic filtration technology to improve the problem of large particles; the accumulation of internal stresses and the decrease of bonding force will be faced when thick films are deposited by arc ion plating, and the poor toughness of the ceramic thick films, continuous grain boundaries, and other defects will cause cracks to arise and expand, resulting in the failure of the ceramic thick films. The poor toughness of ceramic thick films, continuous grain boundaries, and other defects can cause cracks to form and expand, leading to ceramic thick film failure. Multilayers, gradient structures, and special process adjustments can be designed to alleviate internal stresses and bonding to the film base. Huang et al. [13] deposited 13.5 μm TiN/CrN multilayers on the surface of high-speed steel by arc ion plating technique with the growth rate of 60–100 nm/min, the base bonding of monolayer CrN and TiN was 38–45 N, while the base bonding of TiN/CrN multilayers was increased to 55 N. Song46 deposited Cr/CrN multilayers with different modulation ratios by AIP technique, and their hardness and elastic modulus were lower than those of monolayer CrN films by multilayer structure, which can reduce the dislocation motion and thus improve the hardness and fracture toughness of film layers. Song46 deposited Cr/CrN multilayers with different modulation ratios by AIP technique, the hardness and modulus of elasticity were lower than those of single-layer CrN films, and the multilayer structure could reduce the dislocation motion and thus increase the hardness and fracture toughness of the films, but the films started to peel when the film layer reached 10 μm (Fig. 3.5).

3.3 Magnetron Sputtering

Magnetron sputtering technology is a physical vapor deposition technology developed in 1970 and applied in various fields such as aerospace, biomedical instruments, and electronic products. Its principle is shown in Fig. 3.6, the main process of coating is: in the lower working air pressure, the first use of glow discharge to generate plasma, under the action of the electric field ions will get a larger energy, and then bombard the target material of the membrane layer at a faster speed; in the energy-carrying ions under the bombardment of the target material of the atoms to get enough energy to overcome the surrounding atoms of the bondage on it, and from the target material to be dispatched; sputtering out of the membrane layer atoms under the action of orthogonal magnetic field, and the sputtering of the atoms of the target material. The sputtered atoms in the membrane layer under the action of orthogonal magnetic field, and other gas molecules collision, exchange of energy and change the speed, migration to the substrate; sent atoms reach the surface of the substrate after adsorption, condensation, in the surface of the migration and diffusion, collision and nucleation and continuous aggregation of aggregation, and ultimately the growth of the formation of a continuous membrane layer.

Magnetron sputtering has a wide range of applications and is a highly efficient sputtering coating method. At present, the magnetron sputtering process has been applied and studied in the preparation of zirconium alloy cladding coatings, which has many advantages, such as: a wide range of target materials, low sputtering energy, convenient operation of the equipment, controllable quality of the film layer, good reproducibility of the process, controllable deposition rate, and small deformation of the workpiece. Zhong et al. [16] FeCrAl coatings of different compositions were deposited on Zr-2 by magnetron sputtering, and it was found that coatings with an Al content higher than 18 at.% promoted the formation of a thin film of aluminum

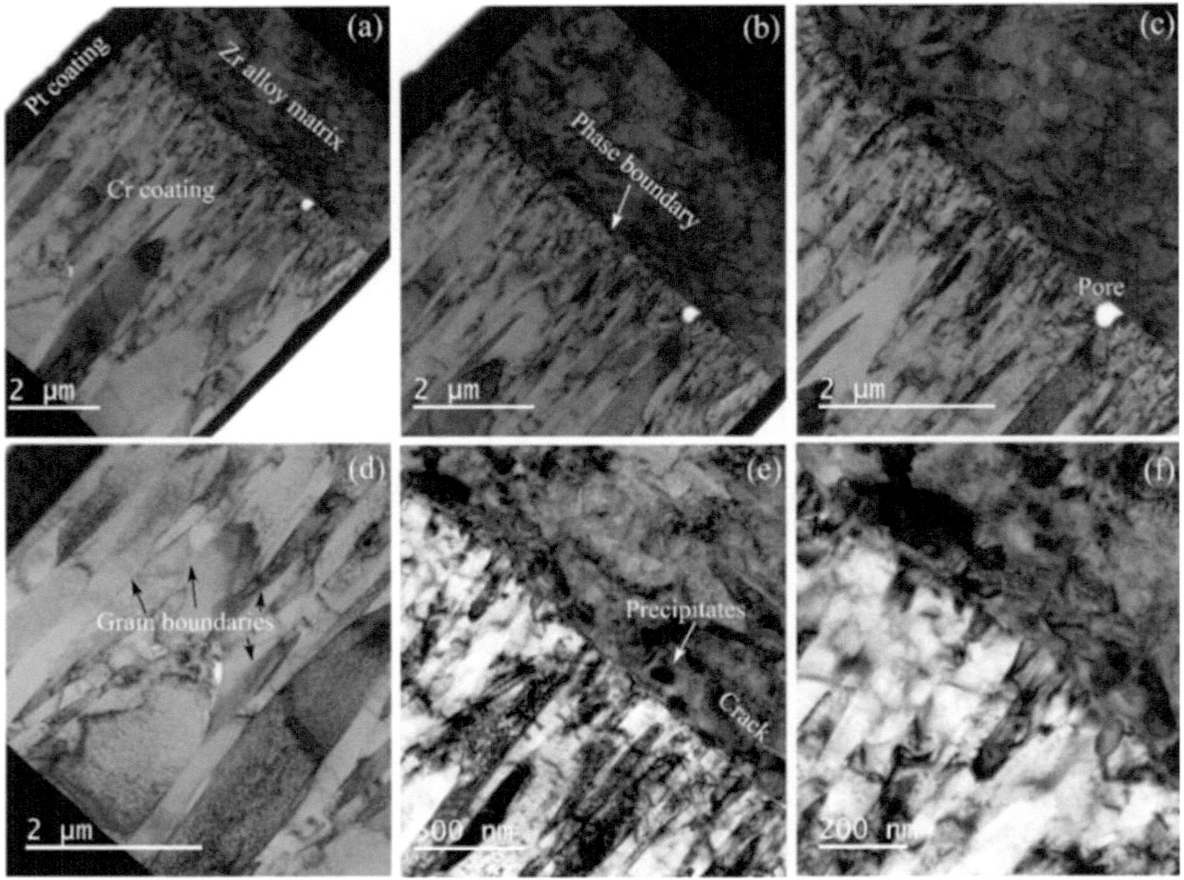

Fig. 3.5 Microstructure magnification of unirradiated Cr-coated Zr-4 alloys progressively increases from (**a–f**) [14]

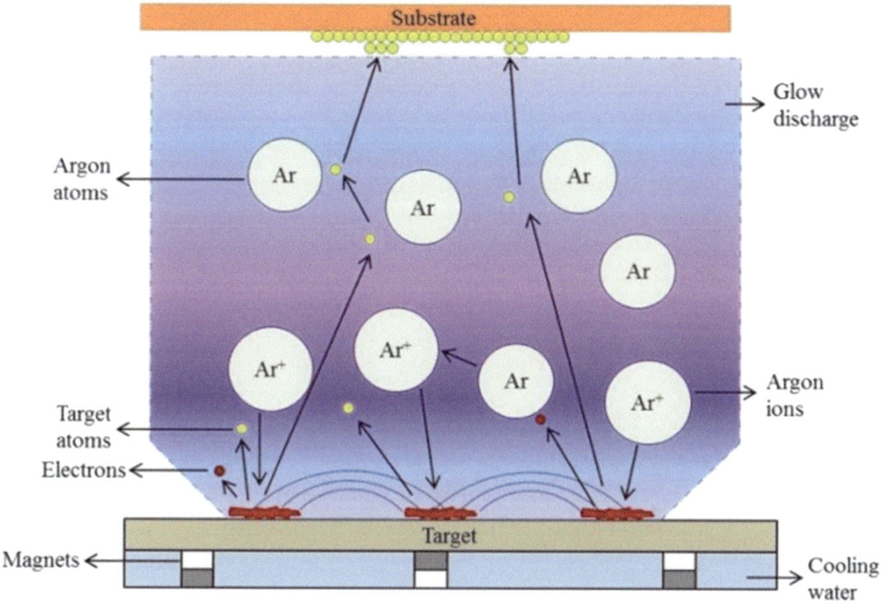

Fig. 3.6 Schematic diagram of the basic principle of magnetron sputtering [15]

oxide at 700 °C vapor, which, in turn, significantly reduced the rate of oxidation of the Zr-2 substrate. D.V. [17] The oxidation properties of Cr coatings deposited on E110 zirconium alloy surfaces by magnetron sputtering and electroplating were compared. The surface of the magnetron deposited Cr coatings was smooth, the microstructure was dense, and the thickness of the coatings was uniform. In contrast, the thickness of the Cr coatings prepared by electroplating is not uniform, and cracks and pores appear underneath the Cr coatings, which affects the bonding of the coating to the substrate as well as the oxidation resistance of the coating. The French AREVA Group has mastered the technology of preparing Cr coatings for full-size zirconium alloy cladding tubes (φ9.5mm, length 3600mm) and has developed a physical vapor deposition equipment capable of preparing Cr coatings for full-size zirconium alloy cladding tubes, as shown in Fig. 3.7.

Magnetron sputtering technology has many advantages, but there are also certain shortcomings: (1) the utilization rate of the target material is low, generally 20% to 30%; (2) followed by the deposition process, in the electrode formed by the action of the uneven magnetic field, the target material is not uniformly etched; (3) the combination of the film and the substrate of the lower strength, the process is more complex. With the continuous research in recent decades, a number of new technologies have been introduced to improve the various deficiencies of magnetron sputtering. At present, the development of new magnetron sputtering technology has multitarget magnetron sputtering technology, pulse magnetron sputtering technology, non-equilibrium magnetron sputtering, and so on. Multitarget magnetron sputtering

Fig. 3.7 PVD equipment for preparation of full-size zirconium alloy clad tube Cr coating preparation [18]

technology, i.e., by regulating the sputtering of multiple targets together, so as to achieve the preparation of composite multielement film layers, Zhang et al. [19, 20] An AlCrMoNbZr high-entropy alloy coating was deposited on N36 alloy by five-target co-patterning, and its morphology was characterized. The structure of the coating was obtained from the analysis as a nanocomposite-type coating with dense organization. The non-equilibrium magnetron sputtering technology is mainly through the use of different permanent magnets in different parts of the target to make the target core and the edge of the magnetic field is different, in order to increase the point ions bypassing, so as to make the uniformity of the prepared coating and the film base bonding strength have been improved compared with the traditional balanced magnetron sputtering, the sputtering range is wider, the target utilization is also higher, and it is commonly used in a variety of industrial and experimental coatings. Figure 3.8 show schematic diagram of a closed-field unbalanced magnetron sputtering system.

High-power pulsed magnetron sputtering technology is to change the traditional DC power supply loading form into a short pulse, high peak power current loading form, which greatly improves the ionization rate of the sputtering target, thus increasing the energy of sputtered ions, and not only is the target utilization rate of the coating prepared using this method high, but the quality of the coating has also been greatly improved. However, the target material in this method has a reabsorption effect, resulting in the deposition rate which is significantly lower than the traditional DC magnetron sputtering technology, and in the case of the same average power of the target material, the deposition rate is lower than the latter. 20–80% [22], Zhou et al. [24] The surface and cross-section morphology of the Cr coatings deposited on Zr-4 alloy by DCMS and HiPIMS were investigated comparatively,

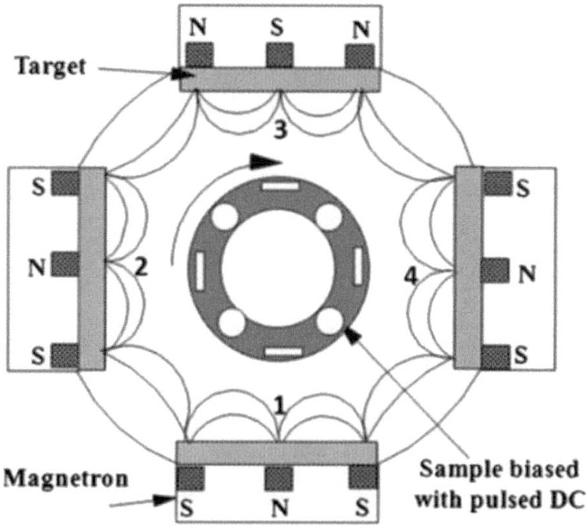

Fig. 3.8 Schematic diagram of closed-field unbalanced magnetron sputtering system [21]

as shown in Fig. 3.9. At the same deposition time, the thickness of the DCMS-Cr coating is 37.5 μm, which is significantly higher than that of the HiPIMS-Cr coating (19.8 μm), indicating that the deposition rate of DCMS is higher. On the other hand, the Cr coatings deposited by DCMS consisted of typical coarse, conical columnar grains, whereas the coatings prepared by HIPIMS showed a more dense structure, and the slower deposition rate limits its application in high-temperature protective coatings.

3.4 Laser Cladding

Laser cladding is a surface modification technology that emerged in the 1970s. It is based on the principle that under the simultaneous irradiation of a laser beam, the surface of the coating and the substrate are rapidly melted and then solidified to form a metallurgically bonded surface coating after the laser beam is removed [26].

Compared with the traditional surface modification technology, laser cladding technology has the characteristics of high energy density, fast heating and cooling, low thermal influence on the substrate, wide selection of cladding powder and controllable dilution rate, etc. At the same time, the cladding layer organization has the characteristics of uniformity and denseness, fewer microscopic defects, and good metallurgical bonding with the substrate with unlimited dimensions, etc. The laser cladding technology can be used to melt alloy elements into the substrate material. As shown in Fig. 3.10, the use of laser cladding technology on the surface of the substrate can achieve the purpose of the alloying element powder into the substrate material.

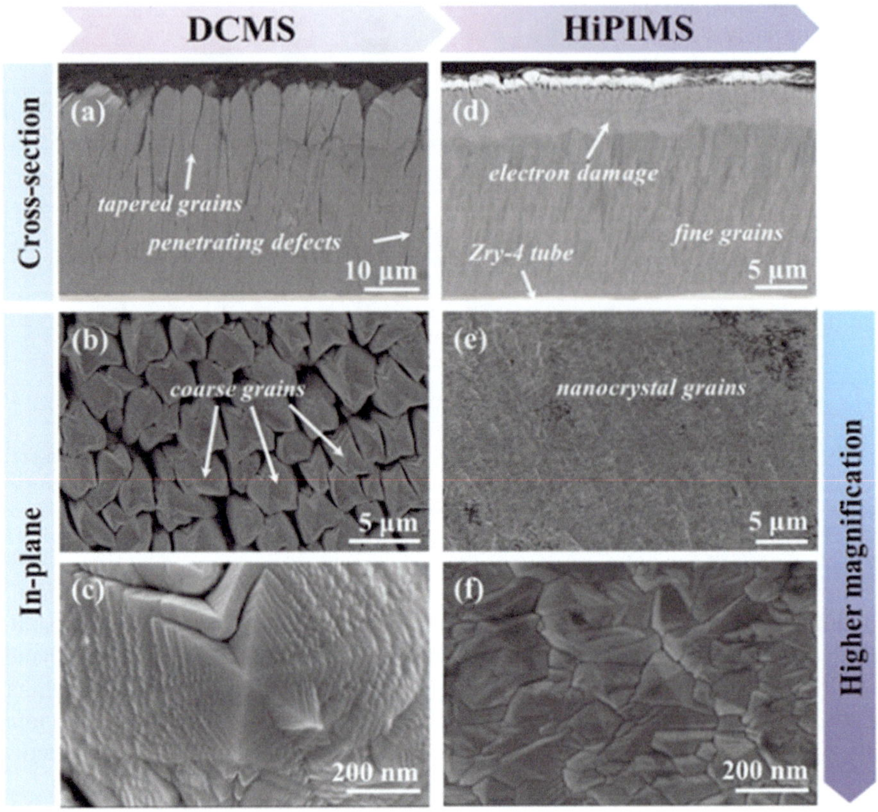

Fig. 3.9 Different cross-section and surface morphology of CMS-Cr and HiPIMS-Cr [25]

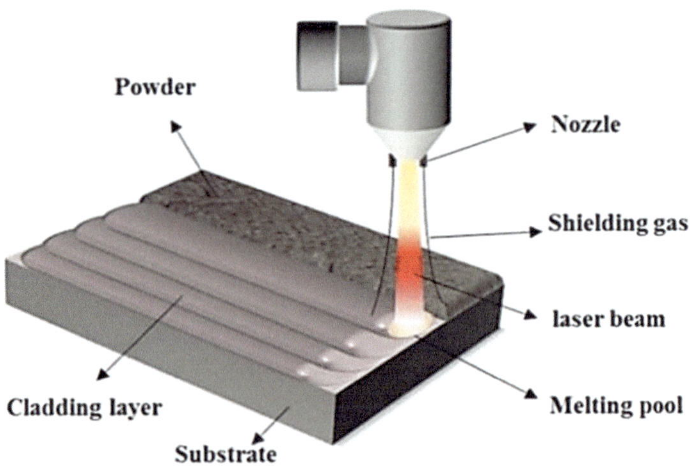

Fig. 3.10 Schematic diagram of laser cladding process

There are two main preparation methods for laser cladding. Two types of powder feeding methods for laser cladding are shown in Fig. 3.11. The synchronous powder feeding method refers to the process where the powder feeder is fed at the same time with the laser after the surface of the substrate has been treated; and the pre-positioned powder method refers to the method of spreading the pre-coated powder uniformly on the surface of the substrate after the surface of the substrate has been treated and then cladding is carried out. The pre-positioned powder method, because the powder is spread evenly on the substrate in advance, is prone to contact with air during the cladding process and form coatings containing defects such as air holes and cracks; however, the simultaneous powder feeding method avoids a large amount of contact between the powder and the air during laser cladding and therefore forms better than the pre-positioned powder and has fewer defects. At the same time, these two methods are also widely used due to the advantages of controllable dilution rate and good bonding strength. The advantage of this new technology is that the primary organisms are uniformly distributed in the solid solution substrate during solidification, and no segregation of alloying elements occurs.

Different process parameters for laser cladding have a large effect on the microstructure of the metal coating, which can change the wear and corrosion resistance of the coating. The solidification process and microstructure of Inconel 718 alloy were investigated under different laser cladding parameters [27]. The results revealed that the microstructure of the cladding layer consisted of columnar crystals and microdendrites. As the laser power increased from 150 to 300 W and the scanning speed increased from 4 to 6 mm/s, the secondary dendrite arm spacing decreased from 1.08 to 0.98 μm. This indicates that higher laser power and faster scanning speeds contribute to finer microstructural features in the cladding layer. In addition, the increase of the feeding speed promoted the formation of equiaxial crystals. Sousa et al. [28] The organization and properties of laser melting-coated Ni–Cr–B–Si alloys on mild steel substrates were investigated, and the results showed

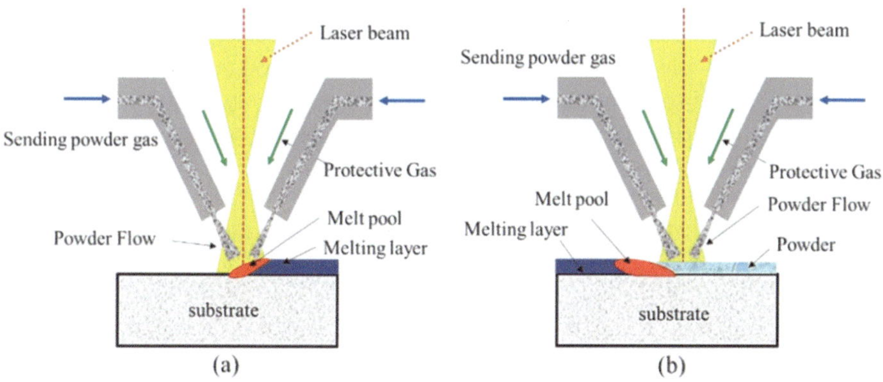

Fig. 3.11 Powder feeding for laser cladding: **a** synchronized powder feeding; **b** pre-positioned powder method

that the coatings at different temperature gradients had different dilution rates and that the coatings with low crack densities and high carbide volume fractions had good wear resistance. Liu et al. [29] Sliding wear experiments on Co_6 coatings on a volume of 30CrNiMo steel at room and high temperatures, respectively, revealed that adhesive wear mainly occurs in high-temperature environments, with good wear resistance and high-temperature oxidation resistance compared to the substrate.

The application of laser melting in the preparation of zirconium alloy cladding coatings is currently limited, mainly because of the difficulty in controlling the coating quality. Laser rapid melting and condensation process, the coating powder due to the heat generated by the gas cannot be discharged in time, residual in the surface to form pores, and pores will damage the uniformity of the fusion cladding layer and is easy to lead to cracks and spalling. At the same time, the rapid heating and cooling of the high-energy laser beam will leave a great thermal stress in the cladding layer, which is also a crack-inducing factor. All these reasons limit the application of laser cladding technology. Therefore, pre-treatment of powders such as heating and drying to reduce gas production, optimization of the operating environment (e.g., development of low-pressure or even vacuum laser cladding) to reduce residual porosity, and improvement of the process (e.g., gradient cladding and pre- and post-heat treatments) to reduce thermal stresses will be the direction of development for the preparation of coatings by laser cladding.

In summary, this chapter discusses a variety of surface coating technologies and their applications in the preparation of zirconium alloy coatings, including spraying, arc ion plating, laser cladding, and magnetron sputtering. The coating deposition rate of spraying is fast and it is easy to prepare thick film, but the disadvantage is high porosity and loose structure; the film adhesion of arc ion plating is high and the film is dense and low porosity, but there are the problems of large particles deposition and non-uniformity of sputtering of coating; the fast cooling and heating of laser cladding technology in the process of preparation will form cracks in the coating, which restricts the application of its zirconium alloy coatings; compared to the other coating preparation technologies, magnetron sputtering technology is more suitable for zirconium alloy coatings because of its targeting technology, which is more suitable for the application of laser cladding. Compared with other coating preparation techniques, magnetron sputtering technology has been widely used in the preparation of zirconium alloy cladding coatings due to its wide range of target material selection, small workpiece deformation, controllable deposition rate, good process repeatability, and many other advantages and has been widely used in industrial applications, so this paper are used to prepare coatings by magnetron sputtering.

References

1. Liu E, Zhang Y, Wang X et al (2019) Tribocorrosion behaviors of thermal spraying WC/Ni_{60} coated 316L stainless steel in artificial seawater. Ind Lubr Tribol 71(6):741–748
2. Duan Z, Yang H, Kano S et al (2018) Oxidation and electrochemical behaviors of Al_2O_3 and ZrO_2 coatingson Zircaloy-2 cladding by thermal spraying. Surf Coat Technol 334:319–327
3. Wang YD, Bai GH, Huang ZB et al (2018) Behavior of plasma sprayed Cr coatings and FeCrAl coatings on Zr fuel cladding under loss-of-coolant accident conditions. Surf Coat Technol 344:141–148
4. Kim HG, Kim IH, Park JY et al (2013) Application of coating technology on Zirconium-basedalloy to decrease high-temperature oxidation; proceedings of the Zirconium in the Nuclear Industry: 17th International Symposium, STP 1543, F
5. Behavior of plasma sprayed Cr coatings and FeCrAl coatings on Zr fuel cladding under loss-of-coolant accident conditions. Surface and Coatings Technology
6. Lee TL, Mi J, Ren S et al (2018) Modelling and neutron diffraction characterization of the interfacial bonding of spray formed dissimilar steels. Acta Mater 155:318–330
7. Alkhimov AP, Kosarev VF, Papyrin AN (1990) A method of cold gas-dynamic deposition. Soviet Physics Doklady 23(1–2):75–76
8. Lee D, Elward B, Brooks P et al (2021) Enhanced flow boiling heat transfer on chromium coated zircaloy-4 Using cold spray technique for accident tolerant fuel (ATF) Materials. Appl Thermal Eng 185
9. Park DJ, Kim HG, Jung YI et al (2016) Behavior of an improved Zr fuel cladding with oxidation resistant coating under loss-of-coolant accident conditions. J Nucl Mater 482:75–82
10. Yeom H, Maier B, Johnson G et al (2018) Development of cold spray process for oxidation-resistantFeCrAl and Mo diffusion barrier coatings on optimized ZIRLOTM. J Nuclear Mater 507:306–315
11. Park JH, Kim HG, Park JY et al (2015) High temperature steam-oxidation behavior of arc ion plated Cr coatings for accident tolerant fuel claddings. Surf Coat Technol 280:256–259
12. Meng MC, Yang L, Wu Y et al (2019) Study of the oxidation behavior of CrN coating on Zr alloy in air. J Nuclear Mater 515:354–369
13. Sun Z, Li Y, Qiu X et al (2023) An in-situ TEM investigation of microstructure evolution in Cr-coated zirconium alloy under heavy ion irradiation: Simultaneous evolution of both Cr coating and Zr-4matrix. Nucl Instrum Methods Phys Res, Sect B 536:97–103
14. Huang MD, Liu Y, Meng FY et al (2013) Thick CrN/TiN multilayers deposited by arc ion plating. Vacuum 89(0):101–104
15. An in-situ TEM investigation of microstructure evolution in Cr-coatedzirconium alloy under heavy ion irradiation: Simultaneous evolution of both Cr coating andZr-4matrix. Nuclear Inst Methods Phys Res Sect B, 536:97–103
16. Maurya D, Sardarinejad A, Alameh K (2014) Recent developments in R.F. magnetron sputtered thin films for ph sensing applications—an overview. Coatings, 756–771
17. Zhong W, Mouche P, Han X et al (2016) Performance of iron-chromium-aluminum alloy surface coatings on Zircaloy 2 under high-temperature steam and normal BWR operating conditions. J Nucl Mater 470:327–338
18. Sidelev Sidelev DV, Poltronieri C, Bestetti M et al (2022) A comparative study on high-temperature air oxidation of Cr-coated E110 zirconium alloy deposited by magnetron sputtering and electroplating. Surf Coat Technol 433:128134
19. Bischoff J, Delafoy C, Vauglin C et al (2018) AREVA NP's enhanced accident-tolerant fueldevelopments: Focus on Cr-coated M5 cladding. Nucl Eng Technol 50:223–228
20. Zhang RT, Yang ZB et al (2018) Preparation, structure, and properties of an AlCrMoNbZr high-entropy alloy coating for accident-tolerant fuel cladding. Surface and Coatings Technol 347:13–19
21. Zhang W, Tang R, Yang ZB et al (2018) Preparation, structure, and properties of high-entropy alloy multilayer coatings for nuclear fuel cladding:A0fcasestudyAICrMoNbZr/(AICrMoNbZr)N. J Nuclear Mater 512:15–24

22. Luo Q, Yang S, Cooke KE (2013) Hybrid HIPIMS and DC magnetron sputtering deposition of TiN coatings: deposition rate, structure and tribological properties. Surf Coat Technol 236:13–21
23. Hovsepian PE, Ehiasarian AP (2019) Six strategies to produce application tailored nanoscale multilayer structured PVD coatings by conventional and high power impulse magnetron sputtering. Thin Solid Films 688:137409
24. Zhou G, Zuo X, Chen R et al (2023) Low-density HiPIMS favored ultrathick and compact chromium coatings on Zircaloy-4 alloys. Phys Scr 98(10):105403
25. Low-density HiPIMS favored ultrathick and compact chromium coatings on Zircaloy-4 alloys. Physica Scripta 98(10):105403 (2023)
26. Li C (2023) Research progress and prospect of laser cladding technology. J Phys: Conf Ser 2419(1):012049
27. Jelvani S, Razavi RS, Barekat M et al (2019) Evaluation of solidification and microstructure in laser cladding Inconel 718 superalloy. Opt Laser Technol 120:105761
28. De Sousa JMS, Ratusznei F, Pereira M et al (2020) Evaluation of solidification and mig deposited by laser cladding process. Tribol Int 143:106002
29. Liu Y, Wu Y, Ma Y et al (2019) High temperature wear performance of laser cladding Co_6 coating on highspeed train brake disc. Appl Surf Sci 481:761–766

Chapter 4
Coordination of Elemental Diffusion and Self-enhanced Oxide Layer on Zr-Si Amorphous Coatings

Abstract The elemental diffusion and corresponding microstructure evolution during oxidation have a direct bearing to excellent oxidation resistance for amorphous silicides. In this work, the oxidation behavior of sputtered amorphous-Zr_xSi_{1-x} ($0.55 \leq x \leq 0.75$) films at subcritical water environment (320°C and 16.4 MPa) was investigated, which can serve as protective coating for nuclear Zr claddings. The oxidation mechanism was determined based on the characterized chemical composition and microstructure of oxide layer. The selective formation of protective Zr–Si–O amorphous oxide scale and Si-rich Zr–Si transition layer was controlled by the changeable atomic Zr/Si ratio and atomic packing density, as proved by AIMD and TEM. EELS and EDS confirm the initial enrichment of Zr in Zr–Si–O, which is conducive to the structural stability. The external diffusion of Zr and the internal diffusion of silicon are suppressed by the gradually formed Si-rich transition layer when x is 0.55. The very low-oxygen permeation in amorphous Zr–Si–O oxide layer and Si-rich transition layer appeared to be beneficial in preventing oxygen permeation to the underlying Zr-4 alloy substrate in 320 °C and 16.4 MPa.

Keywords Phase selection · Oxidation · Coating · Atomic structure · Diffusion

4.1 Introduction

The corrosive failure is one of the main causes that hinder the lifetime of Zr-based nuclear claddings used in pressurized water reactor (PWR) [1, 2]. Zr alloys undergo severe waterside corrosion under normal operating conditions at subcritical environment [3]. The surface formed zirconia layer will undergo phase transition which lead to the formation of cracks and then failed [4]. The lifetime of cladding tubes can be significantly prolonged by coating technology under normal working conditions.

Recently, silicide coatings have been successfully prepared on the surface of superalloys with good oxidation resistance [5]. Yeom et al. also reported that $ZrSi_2$ coating can effectively protect Zircaloy-4 alloy at 700, 1000, and 1200 °C in air condition

J. Hu and G. Cao, *Zirconium Alloy Coatings*,
https://doi.org/10.1007/978-981-96-5028-6_4

[6]. Considering the similar linear thermal expansion coefficient, chemical composition compatibility, excellent mechanical properties, and oxidation resistance, the application of Zr–Si coating on the surface of Zr alloy is feasible [7]. However, more attentions were paid to the protection performance at lose coolant accident condition at high temperature. The stability at high-temperature and high-pressure water was ignored. The ideal coating should not only meet high-temperature oxidation resistance, but also meet relevant requirements under normal operating conditions.

Normally SiO_2 acts as a protective layer and decreases the reaction rate. However, Fournier et al. [8] reported that the solubility of amorphous SiO_2 in pure water increased exponentially with increasing water temperature. Yeom et al. found that the steam corrosion environment led to weight loss of $ZrSi_2$ and resulted in a zirconium oxide scale with silicon depletion [9]. This indicates that too much silicon is unsuitable in the reactor water environment. It is necessary to evaluate the feasibility of Zr_xSi_y coating in Zr-rich range under normal operating conditions in PWRs.

In this work, the fabrication of Zr-Si amorphous films was carried out using high target utilization sputtering (HiTUS) system [10]. Amorphous alloys usually have corrosion resistance one to two orders of magnitude higher than those of polycrystalline alloys with the same composition [11]. The excellent oxidation resistance of amorphous alloys is related not only to their microstructure, but also to the formation of amorphous oxide layer on the surfaces [3, 12], which further inhibits the diffusion of cations and anions in the oxide film. At the same time, amorphous oxides formed on the surface of amorphous alloy have good bonding flexibility and very small stress concentration, which can minimize the mismatch between oxide and matrix [13]. Thence, the formation of a stable amorphous passivation layer and the protective behavior of coatings are of great interest and the key work.

During oxidation, the dominance of cations diffusion or oxygen ion diffusion through the oxide layer depends on microstructure of the oxide and element diffusion in matrix [14]. The diffusion of elements in the matrix was affected by the atomic ratio of Zr/Si in this work. A stable amorphous oxide film was found to grow on uniform $Zr_{55}Si_{45}$ amorphous film with the preferential oxidation of Zr, generating a Si-rich sublayer, which exhibits a highly efficient oxidation blocking effect. A fast establishment of Zr–Si–O amorphous oxide and the Si-rich transition sublayer maintain the stability of amorphous oxide layer. The atomic structure of amorphous film with the capping amorphous oxide was investigated in detail, showing improved adhesion and corrosion resistance, which is excellent for the proposed application.

4.2 Experimental Methods

The Zr–Si coatings were deposited using the S500 HiTUS system. The sources of Zr and Si came from a composite target, based on which the composition of coatings can be controlled. The substrates used in the experiment were Zr alloy and monocrystalline silicon. Before deposition, the substrates and target were cleaned by argon plasma for 2 min. The base pressure before deposition was set to 1×10^{-5} mbar with the argon gas flow rate being maintained at 88.0 sccm. For the fabrication of Zr–Si layer, the plasma launching system (PLS) power and radio frequency (RF) target power were 1200 W and 200 W, respectively. All films were deposited at room temperature without substrate heating to avoid the obvious atomic migration and crystallization. In order to fulfill the compatibility with zirconium alloys, we set zirconium as an enriched element and set the zirconium content in zirconium–silicon amorphous coatings to 55–75%. For convenience, we named the coatings Zr55, Zr65, and Zr75. The oxidation and corrosion behavior of the coatings was studied by exposing them to an oxidizing environment (320 °C and 16.4 MPa). The oxidation test was performed in the static autoclave in deionized water with 0.01 mol/L LiOH solution.

The chemical state of the film surface was characterized using X-ray photoelectron spectroscopy (XPS, AXIS Ultra DLD), with the C1s peak at 284.6 eV serving as a reference for calibration. To remove hydrocarbon contaminants from the surface, the samples were gently cleaned using an argon ion polishing device with an energy of 200 eV. Crystallographic information was obtained via grazing incidence measurements using an X-ray diffractometer (XRD, Rigaku Ultima IV). The surface and cross-sectional morphology of the coatings were examined using scanning electron microscopy (SEM, ZEISS Sigma 500). Microstructural characteristics were analyzed using a field-emission transmission electron microscope (TEM, FEI Tecnai G^2 F20) operated at 200 kV and equipped with an energy-dispersive X-ray (EDX) detector (Oxford XMAX 80 T). Cross-section samples were prepared using a Gatan Model 691 PIPS instrument. TEM cross-sectional imaging of oxide layer thickness was considered a more accurate method for assessing oxidation behavior.

Structures and energies of $Zr_xSi(x = 1/2/3)$ models were computed within density functional theory (DFT). The amorphous structure of the ZrxSi system is obtained by heating the crystal structure to 5000 K and then lowering the temperature. Ab initio molecular dynamics (AIMD) was performed on conventional cells containing around 96 atoms depending on the stoichiometric formula over 90 ps after a pre-equilibrium run for 10 ps, using a time step of 2 fs in the NVT ensembles. The total radial distribution functions of amorphous structure Zr_xSi were calculated by AIMD simulation at 2500 K.

Element migration data were obtained at elevated temperatures from 2000 K up to 3500 K, over affordable simulation time through AIMD calculation [15]. The diffusion coefficients were calculated by correlating them with the average mean square displacements (MSD) obtained from molecular dynamics (MD) simulations conducted over a specific time period t at each temperature. This method enables the quantitative evaluation of diffusion behavior under varying thermal conditions. The diffusion coefficient can be extracted via the equation $D = \frac{1}{2dt} < [\Delta r(t)]^2 >$, where d is the dimensionality factor. Lower temperature data can then be derived by extrapolation from the Arrhenius equation $D = D_0 \exp(-\frac{E_a}{k_B T})$, with D_0 (pre-exponential factor) and Ea (activation energy) as constants for each AIMD system [16].

4.3 Results and Discussions

4.3.1 *Microstructure Evolution of Oxide Layer on Zr_xSi_{1-x}*

Figure 4.1 exhibits a set of TEM images for the Zr75 after oxidation in 320 °C and 16.4 MPa for 20 h. The formed oxide layer is quite thin and compact, of which the thickness is about 20–30 nm. The oxygen content decreases gradually without a clear interface between the oxide and matrix, as shown in Fig. 4.1a. Some nanocrystals were formed in the amorphous oxide layer, which can be indexed as T-ZrO_2 (Fig. 4.1b). The thickness of oxide layer increases to about 50–60 nm after 40 h oxidation in Fig. 4.1c. The high-angle annular dark field (HAADF) image demonstrates that the oxide layer was not smooth. Abnormal T-ZrO_2 growth occurred in some region, as confirmed by the bright-field image and selected area electron diffraction (SAED) pattern in Fig. 4.1d. Based on EDS mapping and point analysis (Fig. 4.1c and d), one can judge that the content of zirconium in the oxide layer was quite high, which was also the cause of large-scale crystallization. The increased zirconium is attributed to the preferential oxidation of Zr. It should be emphasized that creaks and holes were usually formed due to the large volume change of phase transition.

In contrast, the oxide layer of Zr65 demonstrates a better corrosion resistance. The microstructure evolution of the oxidized Zr65 was investigated by TEM and STEM in Fig. 4.2. After 20 h oxidation, the thickness of oxide layer was about 20 nm (Fig. 4.2a). The overall oxygen-deficiency state was identified in the oxide layer during the initial oxidation, which tends to be more obvious to the metal/oxide interface, indicating an oxygen penetration process. The Zr content in the amorphous

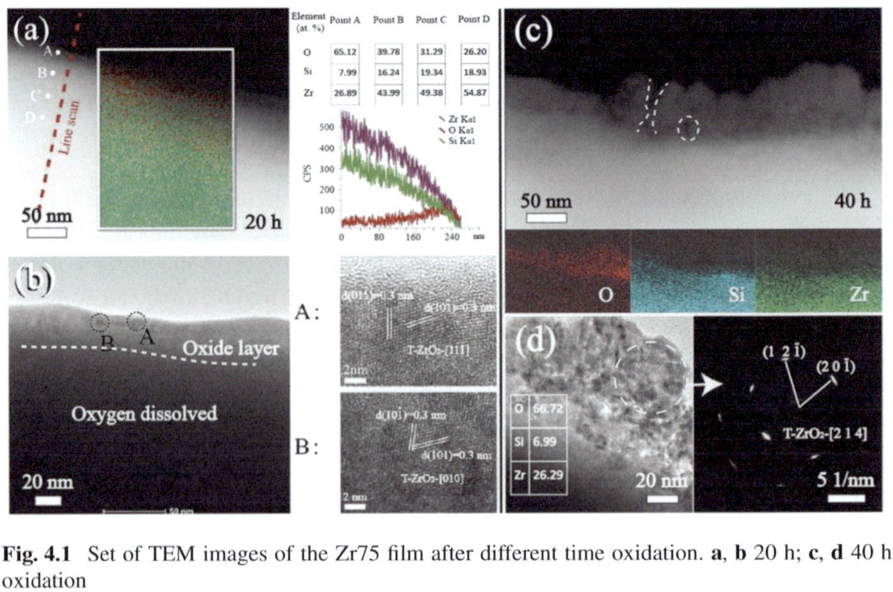

Fig. 4.1 Set of TEM images of the Zr75 film after different time oxidation. **a**, **b** 20 h; **c**, **d** 40 h oxidation

oxide layer was increased after 40 h oxidation. One should also note that because of the large solid solubility of oxygen in Zr [17], there are still more than 14 at.% oxygen beneath the oxide layer (Fig. 4.2c). The oxygen-contaminated region could be as deep as 350 nm in amorphous HfNbTaTiZr alloy, which strongly affects the physical and chemical properties [18]. After oxidation for 100 h in Fig. 4.2e, the oxide scale on Zr65 transformed into a porously crystalline composite structure and peeled off. The abnormal growth and breakage of the oxide layer were related to the gradually increased Zr content and the crystallization induced volume change. The Zr/Si atomic ratio increased from 1.94 after 20 h to 4.73 after 100 h oxidation (Fig. 4.2a and e). Normally the inner diffusion of oxygen is the determining factor of the oxidation process, especially for Zr-based alloys. The cations migration in Zr–Si coatings is related to different oxygen affinity and the chemical potential gradient established in the system. The critical pO_2 for Si to be oxidized is much higher than that for Zr. Therefore, the oxide layer formed in the initial oxidation stage is more based on the internal diffusion of oxygen, and the formed amorphous oxide layer contains more Zr–O bonds. According to literatures, the solubility of silicon in Zr oxides is very limited. If the kinetics permits, silicon will not enter the matrix through the oxide/matrix interface and will preferentially react with oxygen to form zirconium-rich oxides.

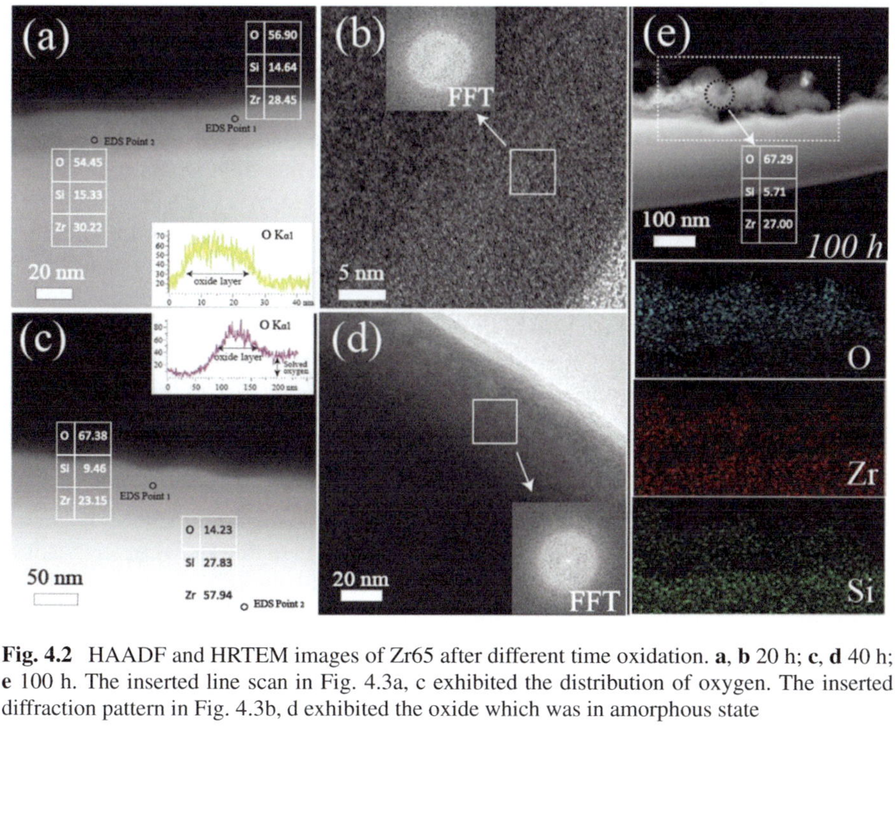

Fig. 4.2 HAADF and HRTEM images of Zr65 after different time oxidation. **a**, **b** 20 h; **c**, **d** 40 h; **e** 100 h. The inserted line scan in Fig. 4.3a, c exhibited the distribution of oxygen. The inserted diffraction pattern in Fig. 4.3b, d exhibited the oxide which was in amorphous state

The chemical and microstructural evolution of the oxide layer directly affects the oxidation resistance. Compared with Zr75, the oxidation resistance of Zr65 was indeed improved, but still suffering from the oxide layer break down and oxygen uptake in matrix after long-term oxidation. In order to further investigate the role of chemical composition, the content of silicon was increased to 45 at.% in this work, namely Zr55. According to the calculated formation energy for Zr–Si amorphous system, Zr55 is more stable than Zr65 and Zr75 [19]. The thickness gain of oxide layer for Zr55 was considerably low and maintained in amorphous state after 30–80 h oxidation, as depicted in Fig. 4.3a–c. The homogeneous amorphous oxide layer always results in improved anticorrosion properties. It is interesting to mention that, with oxidation time increased, a Si-rich layer was formed below the oxide layer. As demonstrated by EDS mapping results in Fig. 4.3c, layer A is an oxide layer (Zr–Si–O) and layer B is a Si-rich (Zr–Si) layer. Actually, this Si-rich layer has begun to form after 30 h oxidation, although the contrast in STEM mode is not obvious due to the small thickness. The two-layer structure can be clearly observed after 50 h oxidation. The Zr/Si atomic ratio of each layer and the matrix was recorded in Fig. 4.3d. With oxidation prolonged, the composition in Si-rich layer tends to be stable (labeled as B). The Zr/Si atomic ratio of the oxide layer was slightly increased (labeled as A). The matrix composition of Zr55 beneath the Si-rich layer kept almost unchanged (labeled as C). It can be tell that after 80 h oxidation, a stable composition was achieved for each layer.

Figure 4.4a exhibits a set of HRTEM images for Zr55 after 80 h oxidation. The Si-rich layer exhibits different contrast from the oxide layer and the matrix. The EDS line scan exhibits a distinct increase in Si content at the reacting alloy/oxide interface and decrease in O content. The HRTEM image and corresponding SAED pattern confirmed that the oxide layer and Si-rich region maintain in amorphous state (a_1,a_2). It is worthy to mention that the interfaces of oxide layer/Si-rich layer/substrate amorphous are quite smooth and compatible since no long-range order structure appears. In order to characterize the bonding and electronic structures of different layers, EELS with nanosized electron probe was introduced, which is sensitive to the chemical environment. In Fig. 4.4c, both silicide spectra have identical features which are dramatically different from both the oxide spectrum. The shift of the threshold to higher energies in Si L-edge reflects the changed chemical state of the absorbing Si atoms in the oxide layer. The O K spectra also demonstrate characteristic peaks in the oxide layer. A comprehensive Zr–O–Si layer with special bonding characters was identified. The onset energy of the Si L-edge is observed to be lower than that in SiO_2, indicating the absence of continuous Si–O bonding in the material. Additionally, the L_3/L_2 ratio increases to 2.51 in the oxide layer compared to that in ZrO_2, suggesting a more significant transfer of electrons from the outer d-band of Zr to oxygen within the Zr–Si–O layer. This highlights distinct electronic interactions in the oxide layer.

The silicon spectra of the two silicides are similar, both displaying two distinct features associated with transitions from the 2p core level to unoccupied states. Peak A corresponds to transitions from 2p to 3s states, while Peak B is attributed to transitions to strongly hybridized 3p states. The p-d-orbital hybridization between Si and Zr results in the transfer of electrons from Zr 4d orbitals to Si 3p states, leading

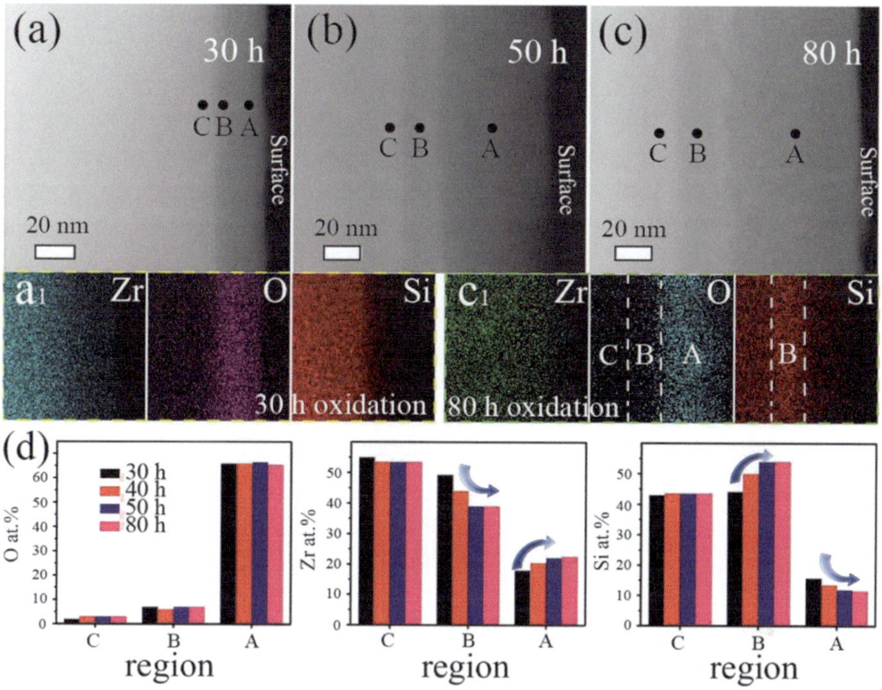

Fig. 4.3 **a–c** Set of HAADF images and EDS mapping results of Zr55 after 30–80 h oxidation: **a** 30 h; **b** 50 h; **c** 80 h. **d** Statistical histogram of composition changes at different locations at different times

to fewer unoccupied states in the Si 3p orbitals. This reflects the strong electronic interaction between Si and Zr in the silicide structure. This was typically enhanced in the Si-rich layer. As can be seen in Fig. 4.4c, the intensity of peak B was obviously decreased for Si-rich layer. The strong covalent bond and the hybridization of the outer-shell valence electrons of Si 3p and Zr 4d enhanced the mutual constraint between the atoms.

For Zr L-edge, all spectra exhibit a prominent feature corresponding to the 2p-4d dipole transitions, commonly referred to as "white lines." Due to the spin–orbit coupling effects in Zr 4d orbitals and the electric dipole transition selection rules, only specific transitions are allowed. At the Zr L_3-edge, the $2p_{3/2} \rightarrow 4d_{5/2}$ transition occurs, while at the Zr L_2-edge, the $2p_{1/2} \rightarrow 4d_{3/2}$ transition is observed. These transitions highlight the electronic structure and spin–orbit interactions of Zr in the material. The L_3/L_2 ratio of Zr55 matrix is slightly lower than that in Si-rich layer, but still higher than the reported 1.87 of metallic Zr. This is attributed to the large electronegativity Si than Zr and the covalent effect, which is leading to an increasing density of empty states with d symmetry above the Fermi level of Zr. The more asymmetric line shape of L_3-edge is also an indicative of the rehybridization of the d orbitals with Si during the Zr–Si bonds' formation.

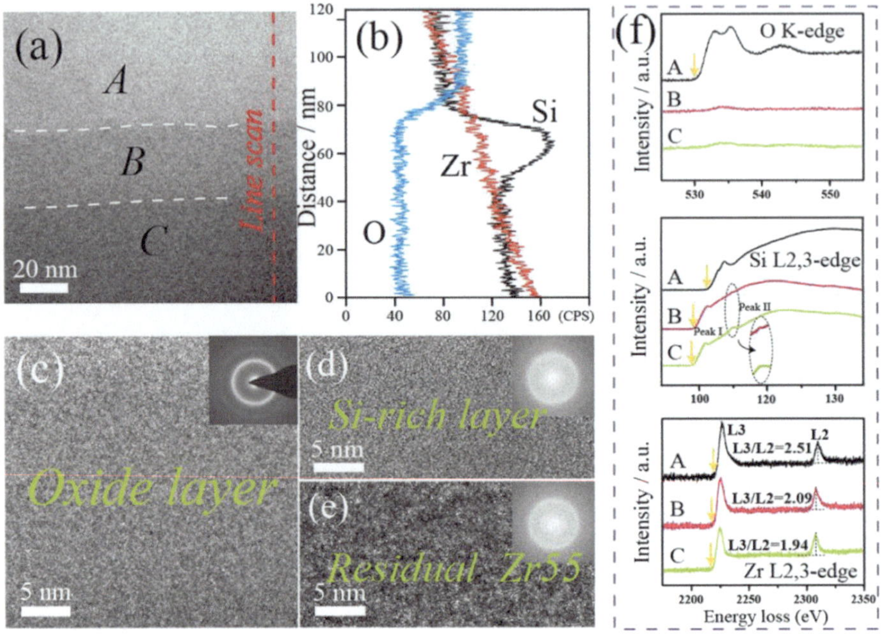

Fig. 4.4 **a** Cross-sectional morphology of Zr55 coating after 80 h oxidation. The bright-field image shows an oxidized layered structure. **b** Elemental profiles of oxidation cross-section marked in Fig. 4.4a. **c** EELS results of different layers

The Formation of Si-Rich Layer. The elemental distribution in the matrix adjacent to the grown oxide layer can deviate from the nominal composition largely, which is attributed to the selective oxidation of active element and oxidation-induced cations diffusion. If the oxidation reaction interface motivates inwardly at a rate significantly faster than that of cations diffusion, the composition of the oxide film is close to that of the matrix. On the other hand, if the reaction interface motivates slowly inward, there will be enough time for cations diffusion. In this case, if the diffusion distance of the cations is limited in the matrix, the high flux cations will accumulate leading to an element-rich layer. For Zr55 in this work, the gradual enrichment of Si beneath the oxide layer was due to the preferential oxidation of Zr from the amorphous matrix and the concurrent oxidation-induced interfacial segregation of Si atoms (analogous to the pile-up of Cu for the Zr–Cu alloys [14] and Cr for NiAlCr alloys [20] under high-temperature oxidation).

To further study the relationship between elemental diffusion and atomic structure in Zr–Si system, simulations based on DFT were carried out. Figure 4.5a demonstrates that the position of the first peak in total radial distribution function (RDF) shifts according to the Si content. The atomic radii of Si are smaller than that of Zr. With the verified Si content, the amorphous atomic packing structure changes. It was reported that in amorphous Al_xZr_y alloys, the increase of Al content leads to a large local packing density [21]. Al and Si are adjacent elements in the periodic table. The electronegativity of silicon is greater than that of aluminum, which brings a better orbital hybridization between zirconium and silicon. Moreover, the covalent atomic radius of silicon (0.111 nm) is smaller than that of aluminum (0.118 nm). Understandably, the increase of Si content contributes to a denser atomic packing of the amorphous structure. Figure 4.5b, c shows that the diffusion activation energies of both oxygen and silicon increase with the increase of silicon content. The diffusion of Si in Zr55 is harder compared with that in Zr65 and a Si-rich layer was piled up in Zr55. For Zr65, no Si-rich layer was formed (Fig. 4.6). This is mainly due to the loose atom packing and high diffusion coefficient of silicon. The back diffused Si atoms can migrate a longer distance, instead of aggregating underneath the oxide layer.

The Role of Si-Rich Layer on the Stabilization of Amorphous Zr–Si–O Oxide. Although amorphous oxide is normally meta-stable, the growth of ultra-thin amorphous oxide layer on metallic alloys can be thermodynamically preferred, which is attributed to the low surface and interface energies [22]. The oxidation behavior of Zr65 and Zr75 amorphous films clearly demonstrated that the amorphous oxide was formed. However, it could not maintain amorphous state with oxidation proceeds. As comparison, the amorphous Zr–Si–O layer on Zr55 coating was quite stable with increased thickness. Once the silicon-rich layer is formed, elemental diffusion becomes difficult. The composition of the oxide layer on the surface is basically stable. The increase of Zr/Si atomic ratio Zr–Si–O layer ceased at approximately 2 with further oxidation, which is important for the amorphous stability. Farabaugh et al. doped SiO_2 in ZrO_2 films and found that when the content of doped silicon

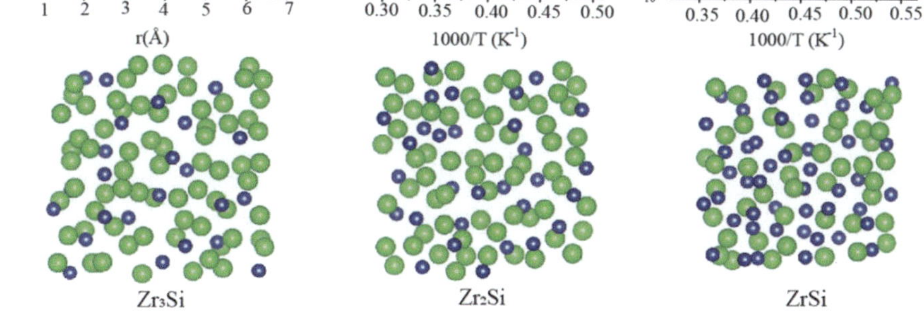

Fig. 4.5 **a** Comparison between total radial distribution functions of am-(Zr, Si) obtained from the AIMD simulations with three different compositions. **b** Diffusion activation energy of oxygen. **c** Diffusion activation energy of silicon

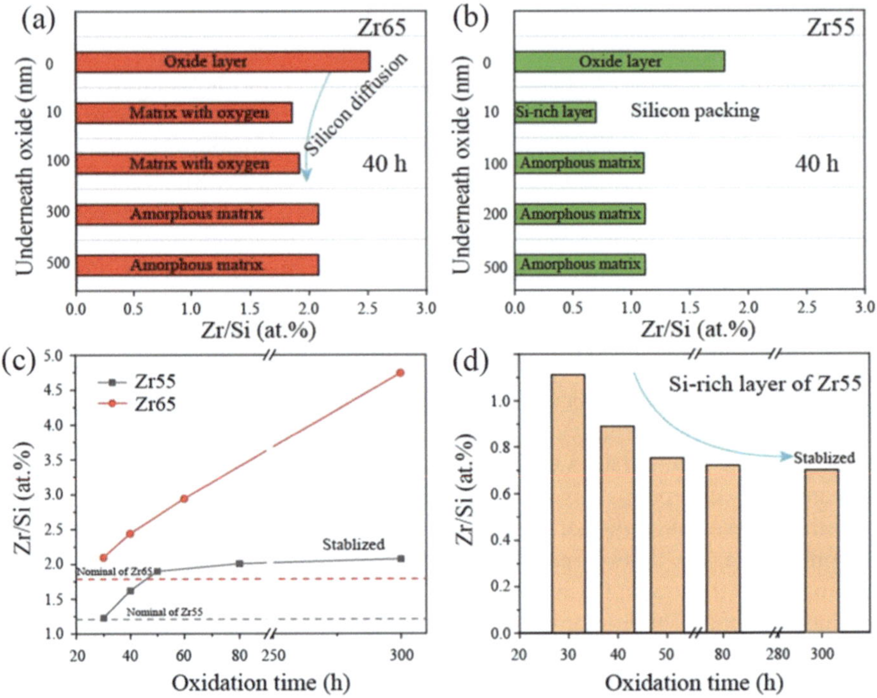

Fig. 4.6 Composition changes for oxide layer and matrix after 40 h oxidation for Zr55 (**a**) and Zr65 (**b**), **c** Zr/Si atomic ratio of oxide layer for Zr65 and Zr55 with oxidation time increased from 30–300 h. **d** Zr/Si atomic ratio of Si-rich layer for Zr55 with oxidation time increased from 30–300 h

oxide exceeds 30 at.%, amorphous film was obtained [23]. The kinetic obstruction is quite large for the formation of crystalline oxide in this Zr–Si–O system.

4.3.2 The Oxygen-Barrier Property and Hydrochemical Stabilization of Zr–Si–O Amorphous Oxide Layer

XPS was carried out to investigate the evolution of chemical states in the oxide layers of Zr55 and Zr65 (Fig. 4.7). No signals of Zr 3d and Si 2p from the pristine metallic Zr–Si glass are found, which confirms the complete oxidation state, and the pristine Zr–Si bond has been totally invaded by oxygen atoms. Figure 4.7a, d shows the deconvolution of O1s after 80h oxidation. The three components corresponding to O-Zr, Si-O-Zr and oxygen vacancies are stabilized near the binding energies of 529.7, 530.6 and 531.8 eV [24], respectively. One could identify that the O-Zr contribution is dominant in the oxide layer of Zr65, and the intensity for Si 2p is relatively weak, indicating a large amount of Zr in the oxide layer. This is consistent with the result in Fig. 4.2. It is also need to mention that oxygen vacancy formed in the oxide film of Zr65 (colored part in Fig. 4.7d), which did not appear in Zr55. Indeed, the oxygen vacancy defects are easily created in transition metal oxides since they are always N-type semiconductors with vast intrinsic point defects. On the other hand, the oxygen is not stable in the oxide layer formed on Zr65, due to the high solid solubility of oxygen in Zr65 matrix. As shown in Figs. 4.1 and 4.2, there exists obviously dissolved oxygen in Zr75 and Zr65 matrix. By continuous dissolution of O from the oxide layer into the alloy substrate, O-vacancy-like defects are easily injected into the oxide layer, analogous to the transport process proposed for the oxidation of Zr metal [25]. Ultimately, the Zr-rich Zr–Si–O layer with oxygen vacancies fails to form an effective barrier against oxygen.

As comparison, the oxygen solubility was reduced quite a lot in Zr55, which is attributed to the increased Si content. The oxygen element in the oxide film does not have sufficient driving force to diffuse into the matrix. The O-vacancy-like defects in Zr–Si–O layer were thus decreased thermodynamically [26]. Moreover, both experiments and calculations have demonstrated that oxygen vacancies in zirconia can trap electrons from the conduction band of Si atoms to form F centers (neutral oxygen vacancies) [27, 28]. The charged oxygen vacancies become uncharged with lower oxygen vacancies mobility [29]. This can also reduce the number of oxygen vacancies in Zr-Si–O layer kinetically. In amorphous oxide, the diffusion of oxygen ion is an anion-"vacancy" exchange mechanism. The decrease in the mobility of vacancies

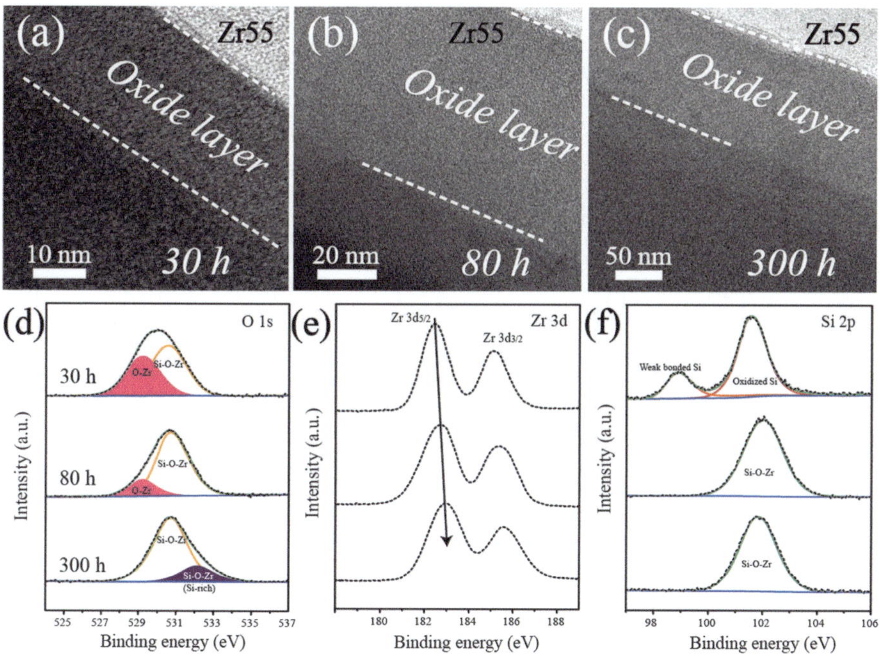

Fig. 4.7 **a–c** XPS results of Zr55 **a–c** and Zr65 **d–f** coating after 80 h oxidation in: O 1s, Si 2p, and Zr 3d. **d** The verified atomic ratio of Zr/Si in oxide layer of Zr55 and Zr65 with oxidation time

can reduce the mobility of oxygen ions and improve the oxygen-barrier ability. It is reasonable to say that there should be enough silicon content in Zr–Si–O layer, either to stabilize the amorphous structure or to reduce the driving force for oxygen to diffuse into matrix.

Although silicon oxide has high protective properties, silicon oxide rapidly transforms into H_2SiO_3 in high-temperature and pressure water. How to coordinate the differential requirements of the stability on materials chemistry and oxygen resistance under working conditions is a key scientific problem in the development of zirconium-based coatings. In this work, no significant dissolution of silicon in Zr–Si–O layer under high-temperature and high-pressure water was observed, as shown in Fig. 4.3f. The hydrochemical stabilization of Zr–Si–O layer on the surface of Zr55 was attributed to the formation of dominant Zr–Si–O bonds (Fig. 4.7a). The binding energy of Zr–Si–O is about 531.8 ± 0.1 eV, which is smaller than that of silica (i.e., 533.4 eV) and larger than that in ZrO_2 (i.e., 529.9). Combined with elemental mapping in Fig. 4.3, it can be concluded that Si and Zr are constituted to a (Si–O-Zr)-type local bonding network in Zr–Si–O amorphous layer. The introduction of ZrO_2 molecule into SiO_2 network is assumed to break the Si–O bonds [30]. If an oxygen atom is bounded to a single Si, it would have unpaired electron in a non-bonding orbital. It is likely that cation (Zr^{4+}) tends to bound with Si–O to form (Si–O–Zr)-type local bonding configuration, since the oxygen affinity of Zr is larger than that of Si. Thus, the (Si–O–Zr)-type local bonding configuration is acceptable [31]. The Zr^{4+} ion has a small d-orbital electron shielding effect, which makes increased bond strength and coating mechanical properties. However, too much Zr content in the oxide layer would lead to crystallization and finally break the oxide passive layer as discussed above. In the XPS results of Fig. 4.7, one can see that after 80 h oxidation, the binding energy of O 1s for Zr55 is higher than that for Zr65. The binding energies shift with different stoichiometries or charge transfer effects. In Fig. 4.7b and e, Zr $3d_{s/2}$ indicates the presence of cationic Zr atoms, which are close to ZrO_2 (182.7 eV). The charge electron transfer of Zr is more extreme in the oxide layer formed on Zr55. Meanwhile, the intensity of Si 2p for Zr55 is also higher than that for Zr65, which means a higher amount of Si the oxide layer formed on Zr55. For Zr65, ZrO_2 begins to nucleate and crystallize due to the higher Zr content. The Si–O bond corresponding to SiO_2 was identified in Zr65 (Fig. 4.7e). The Si-rich layer in Zr55 successfully prevents the excessive gain of Zr in the oxide layer, which maintained the characteristic bonding structure and ensured stability in the oxide layer as well as oxidation resistance.

4.3.3 *Oxidation Kinetic of the Zr55 Coating*

The oxidation kinetics of Zr55, as shown in Fig. 4.8, is compared with that of the Zr-4 alloy. The oxidation process of Zr55 occurs in two distinct stages: Stage I (0 to 60 h) and Stage II (beyond 60 h). During Stage I, the oxide layer thickness increases rapidly, while in Stage II, the oxidation rate slows significantly. The two-stage kinetics has already been reported by Nie et al. for the oxidation of $Zr_{64}Cu_{16}Ni_{10}Al_{10}$ metallic glass [32]. The thermodynamic diffusion of ions through the oxide film and the electron transfer emitted by thermal ions drives the oxidation process. According to Wagner's theory [33], the rate-limiting factor for oxidation is determined by the slow diffusion species, whether positively or negatively charged. In Stage I, the high oxidation rate is primarily attributed to the interface reactions and the accumulation of an immature oxide phase, which facilitates a fast ion migration rate. Conversely, in Stage II, the growth rate of the oxide layer decreases significantly due to the formation of a dense oxide layer and a Si-rich layer, which act as barriers, reducing ion diffusion and slowing the oxidation process.

In Zr55, a closely packed atomic structure of amorphous oxide layer was formed with the Zr/Si atomic ratio about 2. This Si-contained oxide layer proved to be inert in aqueous environment. The initial increase of Zr in the oxide layer could bring a strengthening effect. Unlike Zr65 and Zr75, the stabilization of atomic content in oxide layer of Zr55 is derived from the interior atomic packing property and the formation of Si-rich layer. Thus, the oxidation kinetics of the amorphous Zr–Si alloys is partially governed by the Si segregation beneath the reacting oxide/alloy interface. Figure 4.9 demonstrates the schematic diagram of Zr–Si oxidation mechanism. The barrier effect of ZSO and Si-rich layer would bring slow oxidation kinetics.

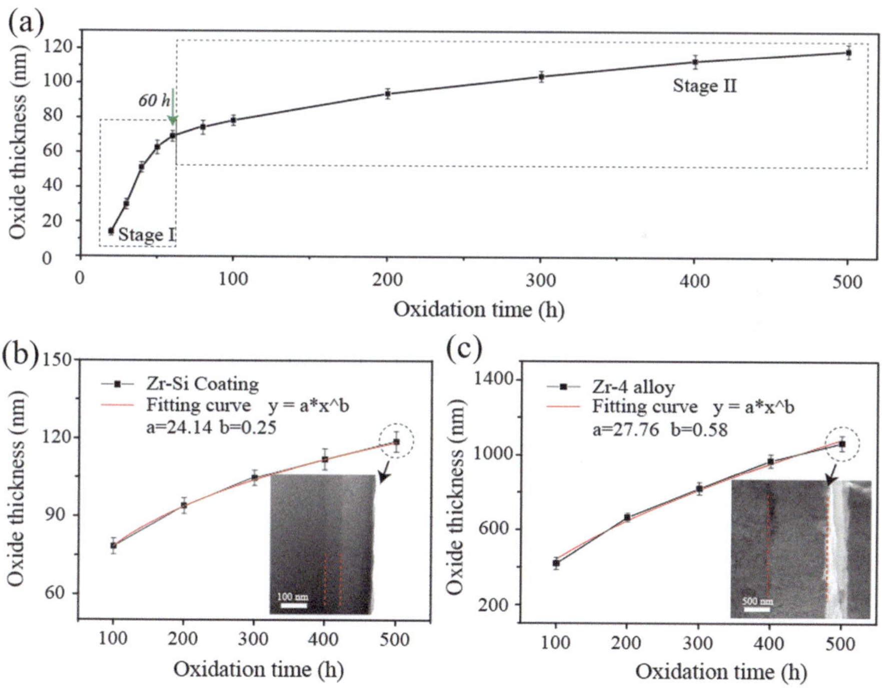

Fig. 4.8 Oxide thickness of the Zr-4 alloy with and without Zr55 coatings as a function of oxidation time

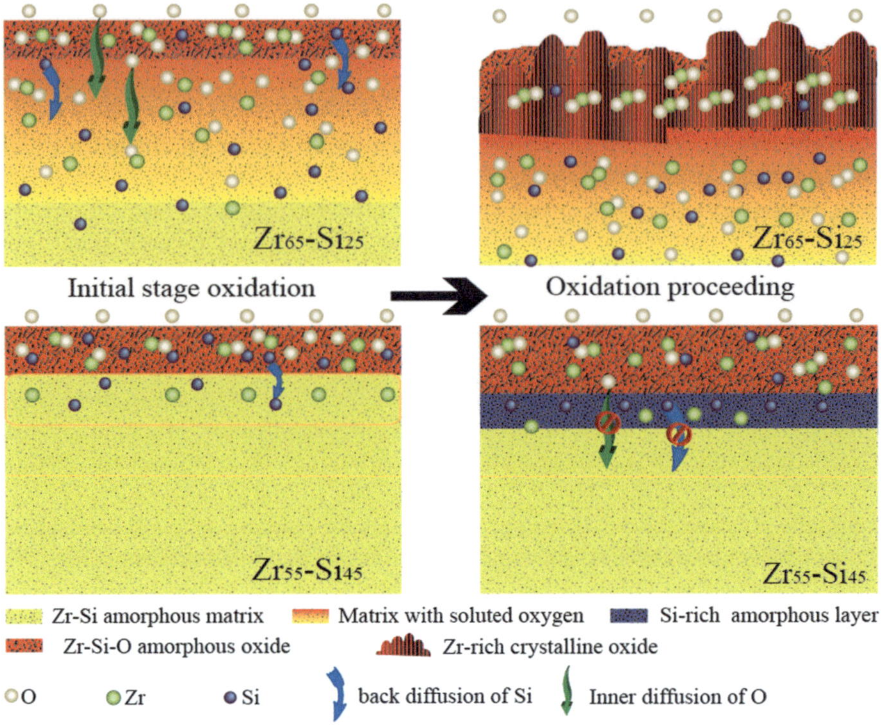

Fig. 4.9 Schematic diagram of Zr–Si oxidation mechanism

4.4 Conclusions

The oxidation mechanism of Zr_xSi_y amorphous films was discussed in this work. A two-layered structure with uniform thickness was observed for $Zr_{55}Si_{45}$, indicating superior oxidation resistance in LiOH-containing water at 320 and 16.4 MPa. In the initial stage, an amorphous (Zr, Si) oxide layer is formed by the inward diffusion of oxygen, and the initial increase of Zr in the oxide layer will form a bond strengthening effect. The initial increase of Zr in the oxide layer could bring a bond-strengthening effect. The cations' diffusion and segregation directly behind the oxidation frontier determine the stable oxidation stage. Based on AIMD calculation, the atomic packing density strongly affects the elements diffusion and the sublayer formation. There are two oxidation kinetics-limiting factors. One is the amorphous oxide layer which exhibits distinct short-range order with Zr–Si–O atomic network structure, preventing silicon from dissolving in hydrochemical environment. The oxidation resistance was also codetermined by the formation of Si-rich sublayer at the reacting interface, which blocks the elemental diffusion and stabilized amorphous oxide layer.

References

1. Zinkle SJ, Was GS (2013) Materials challenges in nuclear energy. Acta Mater 61:735–758
2. Cao G, Ren Y, Wu K, Yao H, Hu J, Shao G, Yuan G (2019) Life extension strategy and research progress of zirconium alloy cladding applied to light water reactors. Surf. Technol. 48:69–81
3. Revesz AG, Fehlner FP (1981) The role of noncrystalline films in the oxidation and corrosion of metals. Oxid Met 15(3–4):297–321
4. Motta T, Couet A, Comstock RJ (2015) Corrosion of zirconium alloys used for nuclear fuel cladding. Annu Rev Mater Res 45:311–343
5. Mitoraj-Królikowska M, Godlewska E (2018) Silicide coatings on Ti-6Al-1Mn (at.%) alloy and their oxidation resistance. Surface Coatings Technol 334:491–499
6. Yeom H, Maier B, Mariani R, Bai D, Fronek S, Peng X, Sridharan K (2017) Magnetron sputter deposition of zirconium-silicide coating for mitigating high temperature oxidation of zirconium-alloy. Surf Coat Technol 316:30–38
7. Kim M, Noh H, Cheol Lee G, Yeom H, Kim TK, Man Kim J, Kim TH, Jin Jo H, Sun Park H, Sridharan K, Kim MH (2022) Flow boiling critical heat flux enhancement in $ZrSi_2$ accident-tolerant fuel cladding with porous structures. Appl. Thermal Eng 207:118164
8. Fournier R, Rowe J (1977) The solubility of amorphous silica in water at high temperatures and high pressures. Am Mineral 62:1052–1056
9. Yeom H, Lockhart C, Mariani R, Peng X, Corradini M, Sridharan K (2018) Evaluation of steam corrosion and water quenching behavior of zirconium-silicide coated LWR fuel claddings. J Nucl Mater 499:256–267
10. Lofaj F, Kvetková L, Hviščováa P, Gregor M, Ferdinandy M (2016) Reactive processes in the high target utilization sputtering (HiTUS) W-C based coatings. J Eur Ceram Soc 36(12):3029–3040
11. Gebert A, Gostin PF, Schultz L (2010) Effect of surface finishing of a Zr-based bulk metallic glass on its corrosion behavior. Corros Sci 52(5):1711–1720
12. De Souza RA, Pietrowski MJ, Anselmi-Tamburini U et al (2008) Oxygen diffusion in nanocrystalline yttria-stabilized zirconia: the effect of grain boundaries. Phys Chem Chem Phys 10(15):2067–2072
13. Weller K, Wang Z, Jeurgens LPH et al (2015) Thermodynamics controls amorphous oxide formation: Exclusive formation of a stoichiometric amorphous $(Al_{0.33}Zr_{0.67})O_{1.83}$ phase upon thermal oxidation of A-Zr. Acta Materialia 94:134–142
14. Lim KR, Park JM, Park SH et al (2014) Oxidation induced amorphous stabilization of the subsurface region in Zr-Cu metallic glass. Appl Phys Lett 104(3):031604
15. Xu H, Xiao W, Wang Z, Hu J, Shao G (2021) J Energy Chem 59:229–241. (2019) ACS Appl. Energy Mater 2:6288–6294
16. Agnihotri MV, Chen SH, Beck C, Singer SJ (2014) J Phys Chem B 118:8170–8178
17. Holmberg B, Dagerhamn T (1961) X-ray studies on solid solutions of oxygen in a-zirconium. Acta Chem Scand 15(4):14
18. Oxidation of amorphous HfNbTaTiZr high entropy alloy thin films prepared by DC magnetron sputtering. J Alloys Compounds 869:157978 (2021)
19. Shim JY, Kwak JS, Chi EJ et al (1995) Formation of amorphous and crystalline phases, and phase transition by solid-state reaction in Zr/Si multilayer thin films. Thin Solid Films 269(1–2):102–107
20. Nijdam TJ, Jeurgens LPH, Sloof WG (2003) Modelling the thermal oxidation of ternary alloys—compositional changes in the alloy and the development of oxide phases. Acta Mater 51(18):5295–5307
21. Weller K, Zotov N, Wang ZM et al (2015) Atomic structure, electronic structure and thermal stability of amorphous Al_xZr_{1-x} $(0.26 \le x \le 0.75)$. J Non-Crystalline Sol 427:104–113
22. Reichel F, Jeurgens LPH, Mittemeijer EJ (2008) The thermodynamic stability of amorphous oxide overgrowths on metals. Acta Mater 56(3):659–674

23. Farabaugh EN, Feldman A, Sun J et al (1987) Examination of thin films in the ZrO_2–SiO_2 system by transmission electron microscopy and x-ray diffraction techniques. J Vac Sci Technol, A: Vac, Surf Films 5(4):1671–1674
24. Donley MS, Baer DR, Stoebe TG (1988) Nitrogen 1s charge referencing for Si_3N_4 and related compounds. Surf Interface Anal 11(6–7):335–340
25. Monclús MA, Callisti M, Polcar T et al (2017) Selective oxidation-induced strengthening of Zr/Nb nanoscale multilayers. Acta Mater 122:1–10
26. Wang PC, Lee JW, Yang YC et al (2016) Effects of silicon contents on the characteristics of Zr-Ti-Si-W thin film metallic glasses. Thin Solid Films 618:28–35
27. Youssef M, Yildiz B (2012) Intrinsic point-defect equilibria in tetragonal ZrO_2: density functional theory analysis with finite-temperature effects. Phys Rev B 86:144109
28. Foster AS, Sulimov VB, Gejo FL, Shluger AL, Nieminen RM (2001) Structure and electrical levels of point defects in monoclinic zirconia. Phys Rev B 64:224108
29. Eichler (2001) Tetragonal Y-doped zirconia: structure and ion conductivity. Phys Rev B 64:174103
30. Lucovsky G, Rayner GB Jr (2000) Microscopic model for enhanced dielectric constants in low concentration SiO_2-rich noncrystalline Zr and Hf silicate alloys. Appl Phys Lett 77(18):2912–2914
31. Rashkeev SN, Glazoff MV, Tokuhiro A (2014) Ultra-high temperature steam corrosion of complex silicates for nuclear applications: a computational study. J Nuclear Mater 444(1–3):56–64
32. Nie XP, Yang XH, Chen LY et al (2011) The effect of oxidation on the corrosion resistance and mechanical properties of a Zr-based metallic glass. Corros Sci 53(11):3557–3565
33. Young DJ (2011) Predicting internal oxidation: building on the Wagner model. MSF 696:1–11

Chapter 5
Interface Effect on the Oxygen Barrier Under Sub-critical and Atmospheric Environment of Zr-Si Multilayer Systems

Abstract The stabilization of interface plays a vital role for the application of multilayer system used in protection of cladding materials. Zr/Si multilayer coating was designed for this strategic application. Comprehensive characterization revealed the structural and phase evolution during oxidation. The amorphous Zr–Si–O (ZSO) layer was formed in situ on the Zr/Si interface during the oxidation process. A competitive consumption of the top Zr layer in silicidation and oxidation was proposed to explain the formation of ZSO. The Zr/Si multilayers with top Zr layer prevent the dissolution of Si and demonstrate better corrosion performance with the formation of ZSO.

Keywords Zirconium · Sputtered films · TEM · STEM · Interfaces · Oxidation

5.1 Introduction

Zirconium alloy fuel claddings have been investigated extensively in pressurized water reactors (PWRs) for several decades [1–3]. However, a serious degradation of Zr alloy may occur due to the harsh and complicated environment in the reactor core (coolant, stress, vibration, irradiation, and gradients in temperature) that can damage the performance of the cladding materials [4–7]. Hence, significant efforts have been made to improve the stability of fuel claddings. Among them, the protective coatings on Zr alloy cladding have been considered as the most effective way to enhance corrosion resistance [8–15].

When designing coatings for Zr alloys, physical and chemical properties must be considered simultaneously, such as low neutron absorption cross-section, radiation resistance, oxidation resistance, tritium permeability, thermal conductivity, and stability in hot water chemical environments. It has been reported that multilayer structures with larger volume fraction interfaces have higher strength and stronger radiation damage tolerance compared to single-layer materials [16, 17]. According to reports, interfaces can act as sliding barriers and absorb defects caused by radiation, such as vacancies and gaps [18, 19]. In the application environment, the oxidation

J. Hu and G. Cao, *Zirconium Alloy Coatings*,
https://doi.org/10.1007/978-981-96-5028-6_5

performance is also important. Thus it is very necessary to research the oxidation behavior of multilayer systems and discuss the unique interface effect for oxidation resistance. Theoretically, the interface would also endow extraordinary advantages to the anticorrosion property, such as a barrier to the progression of cracks, inward diffusion of oxygen, and cross-migration of metal ions [20–22]. Some researches on multilayer coatings have been performed to analyze the oxidation resistance, such as TiAlN/VN and TiAlSiN/CrAlYN [23–26]. Common problems can be summarized is that the element interdiffusion, phase decomposition, and formation of columnar grains in the crystalline layer could strongly decrease the stability of multilayer system. Undoubtedly, chemical and morphological stability of the interfaces is a crucial factor in the design of multilayers. To the best of our knowledge, the antioxidation mechanism based on interfacial effect has not yet been demonstrated in detail.

The multilayer structure can display the advantages of each component. It is very important to find suitable elements for a multilayer system. Considering the good compatibility between coatings and substrates [27], introducing zirconium into multilayer films is of great significance. In addition, silicon with a very low neutron absorption cross-section has lower solid solubility and diffusion coefficient for oxygen [28, 29]. Silicon-based amorphous coatings have extremely high oxidation resistance up to 1400 °C in air [30], which attribute to the formation of dense SiO_2 layer. Cr has excellent high-temperature oxidation resistance and can be used as a protective coating for nuclear cladding materials [31–33]. Due to its excellent corrosion resistance, nickel is also an additive element in zirconium alloys [34, 35]. In this work, non-metallic and metallic elements were selected for the first time to design a multilayer system of Zr/X (X is Ni, Cr, Si) structure. In order to highlight the interface characteristics of multilayer thin films, the zirconium layer was set as the top layer that is easy to oxidize into zirconia phase. The zirconia layer formed on the surface can serve as a barrier for SiO_2 to react with hot H_2O under normal operating conditions, i.e., high-temperature and high-pressure water [36–39].

This study attempts to develop a new multilayer system via sputtering to understand the relationship between the oxygen-barrier property and microstructure evolution in the multilayer structure. Different oxidation behavior of Zr/X multilayer coatings was identified and analyzed to explore the multilayer system with stable structure and excellent oxidation resistance during normal operating conditions in light-water reactors. A unique interface effect of Zr/Si multilayer system was analyzed in this study. The antioxidation mechanism of Zr/Si multilayer is discussed.

5.2 Experimental Methods

Multi-layer deposition was carried out using S500 (Plasma Quest Ltd, UK) technology in an Ar environment. Zirconium (99.999%) and silicon (99.99%) were used as sputtering targets. Before deposition, the substrate was cleaned with ultrasonic equipment in acetone and ethanol for 20 min and then dried it with a nitrogen gun.

The substrate rotates at a speed of 30 rounds per minute to improve the uniformity of the multilayer film. When the vacuum degree of the chamber for preparing multilayer coatings reaches 8.0 ~ 9.8 × 10^{-6} mbar, argon gas was injected into the chamber, and the working pressure during the coating deposition process stabilizes at around 5.3 × 10^{-3} mbar. For the manufacturing of Zr layer, the power of plasma emission system (PLS) and direct current (DC) target was 1000 W and 250 W, respectively. As for the silicon layer, the PLS power and radio frequency (RF) target power were 800 W and 200 W, respectively. For oxide layer deposition, the oxygen flow rate was fixed at 8 sccm. When the PLS power and DC target power were 800 W and 200 W, respectively. The sputtering rate was determined by a Dektak XT stepper (probe surface profiler). Oxidation experiments were conducted in an industrial furnace at 400 °C in air. The Zr/Si multilayers were also tested for different durations in 320 °C and 15.5 MPa in a static autoclave of a limited company.

The cross-sectional microstructure was characterized by transmission electron microscopy (TEM, FEI Tecnai G^2 F20) and energy-dispersive X-ray spectroscopy (EDS, Oxford XMAX 80T). All element ratios in this study are atomic ratios. Cross-sectional samples were prepared by ion beam thinning using 4keV Ar ions on Gantan 691. The thickness variation was characterized by cross-sectional samples prepared, and the average value was calculated after measuring 10 different samples and regions. The surface morphology of multilayer films was characterized by field-emission scanning electron microscopy (FE-SEM).

5.3 Results and Discussion

Figure 5.1 shows the macrostructure of Zr/Ni multilayer after oxidation at 400 °C for 60 h. The layered structure has almost collapsed. In Fig. 5.1a, one can see that the outermost Zr layer was completely oxidized, beneath which, all other Zr layers contain a large amount of Ni. There are reports that Ni rapidly diffuses into the Zr layer with solid-state reaction, resulting in an increase in the thickness of the Zr layer. The formation of transverse pores in the nickel layer is attributed to the Kirkendall effect in the crystal diffusion couple. In Fig. 5.1b, the lack of mutual diffusion at the ZrO_2/Ni interface indicates that Ni element will not diffuse into the Zr layer containing a large amount of O element. In other words, Ni atoms will diffuse into the Zr layer under low-oxygen partial pressure. Similar phenomena also occur during the oxidation process. It can be concluded that the structure of Zr/Ni multilayer films is unstable due to the severe diffusion behavior dominated by one species (Ni).

Figure 5.2 shows the microstructure of Zr/Cr multilayers deposited and oxidized for 60 h. It can be noted that the deposited metal layer contains nanocolumnar grains, especially for the Cr layer. After oxidation, the Cr layer still contains obvious columnar crystals, while the Zr layer is no longer obvious. The EDS line-scan results are shown in Fig. 5.2c. Although the Cr layer was not oxidized, the buried Zr layer was completely oxidized, which means that the Cr layer failed to prevent the internal

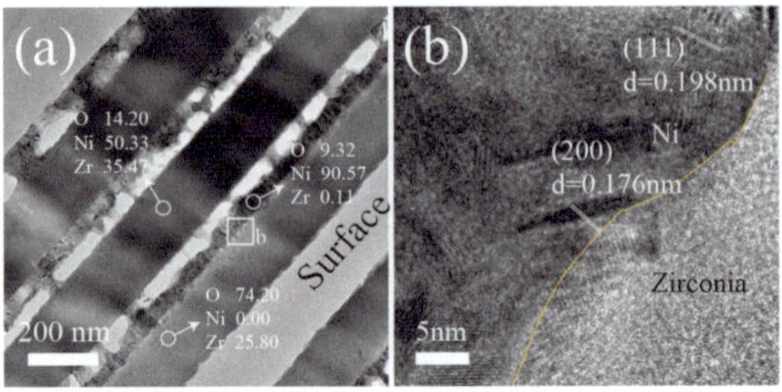

Fig. 5.1 Set of TEM images and composition analysis of the Zr/Ni multilayer film after oxidation at 400 °C for 60 h in air. **a** Bright-field image of the cross-sectional morphology with EDS result shown as inset **b** HRTEM image of the interface of Ni and ZrO_2 layer

diffusion of oxygen. The thickness increase of the Zr layer was due to the absorption of oxygen.

Through the above results, it can be confirmed that although the Cr and Ni layers are not completely oxidized, they cannot prevent the diffusion of oxygen and cannot guarantee the stability of the structure. Two key factors need to be considered in the design of antioxidation and structurally stable multilayer films: (1) dense and stable oxygen-barrier layer; (2) suitable mutual diffusion coefficient with zirconium.

As shown in Table 5.1, the diffusion coefficient of silicon in zirconium is much smaller than that of nickel. Moreover, the amorphous properties of the silicon layer can avoid the formation of defect channels such as grain boundaries.

Figure 5.3 shows the cross-sectional morphology of the prepared Zr/Si multilayer with clear interface. A Zr-Si amorphous layer was sputtered as a buffer layer to obtain high interface quality and good adhesion of the Zr/Si multilayers on the substrate. The layer thickness of Zr and Si was 35 nm and 25 nm, respectively. The high-resolution TEM images of the Zr/Si layer in Fig. 5.3b show the Zr layer composed of nanocrystalline particles, while the Si layer is in the amorphous state. The crystalline

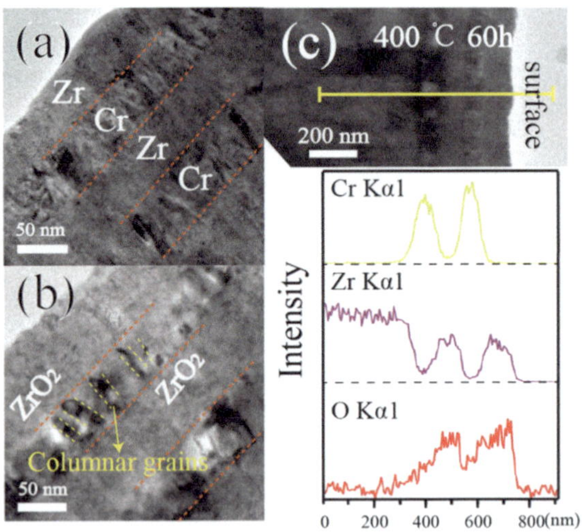

Fig. 5.2 Microstructure and composition scanning of Zr/Cr multilayer deposited and oxidized at 400 °C for 60 h: **a** cross-section of as-deposited multilayer, **b** cross-section for 60 h, **c** composition scanning for 60 h

Table 5.1 Key performance of elements

Element	The diffusion coefficient (in Zr)
Si	1.26×10^{-9} cm^2/s (840 °C)
Cr	2.8×10^{-9} cm^2/s (830 °C)
Ni	8.0×10^{-7} cm^2/s (830 °C)

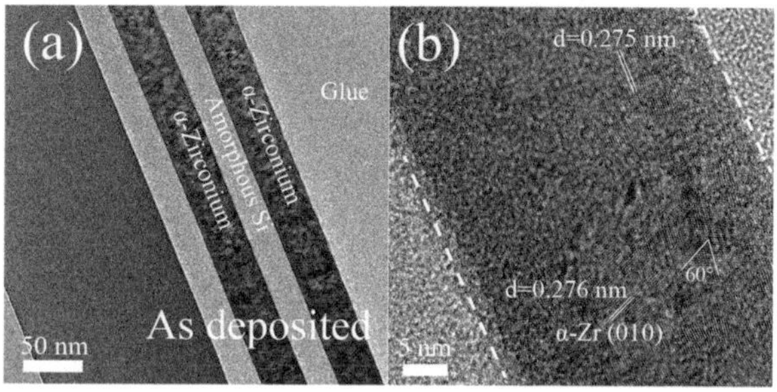

Fig. 5.3 Cross-sectional morphology of the as-prepared Zr/Si multilayer

size in the Zr layer is about 20–30 nm. The multilayers display a dense microstructure without cracks or spallation.

After oxidization at 400 °C for 10–200 h in ambient air, Fig. 5.4 shows that multilayer structure maintained stable. The thickness of top and the second Zr layers increased to ~ 50 and ~ 40 nm, respectively. Element scanning demonstrates that the thickness gain in the top and second Zr layer can be attributed to oxidation and silicidation process, respectively. The top Zr layer was completely oxidized as the Zr/O ratio is approximately 1:2 after 10 h oxidation (Fig. 5.4a, c). Both tetragonal ZrO_2 (T-ZrO_2) and monoclinic ZrO_2 (M-ZrO_2) were detected by SAED pattern (Fig. 5.4a). As usually reported, the T-ZrO_2 transformed into M-ZrO_2 after 200 h exposure, resulting in a single-phase ZrO_2 layer. HRTEM and EDS results of ZrO_2/Si interface (Fig. 5.4b, e) confirm the formation of Zr–Si–O layer (abbr.: ZSO), which is in amorphous state (Fig. 5.4e). The Zr/Si ratio keeps constant of 1 and oxygen content increased from 48.49 at. % (2 h) to 68.33 at. % (200 h) in ZSO during oxidation. The remained second Zr layer was slightly incorporated by trace O. On Zr/Si interface, the diffusion of Si in to Zr results in the formation of amorphous silicide phase on the both sides of the second Zr layer (Fig. 5.5).

Role of α-Si Layer: The critical oxygen partial pressure for the formation of ZrO_2 is quite low which cannot activate the oxidation of Si at the *ZrO_2/Si* interface. At the same time, due to the amorphous structure of the silicon layer, it is difficult for oxygen ions to diffuse inward without fast channels, such as a grain boundary. As an amorphous structure, how about the protective effect of silicon oxide coating? An amorphous SiO_x layer was introduced and a *Zr/SiO_x/Zr* system was fabricated and oxidized in this work. Figure 5.6a, b exhibits the bright-field image of this multilayer before and after oxidation at 400 °C for 60 h. The composition was measured by EDS in atomic ratio. After 60 h of oxidation, the Zr layer on both the sides of SiO_x was fully oxidized. This indicates that SiO_x layer fails to act as an oxygen barrier.

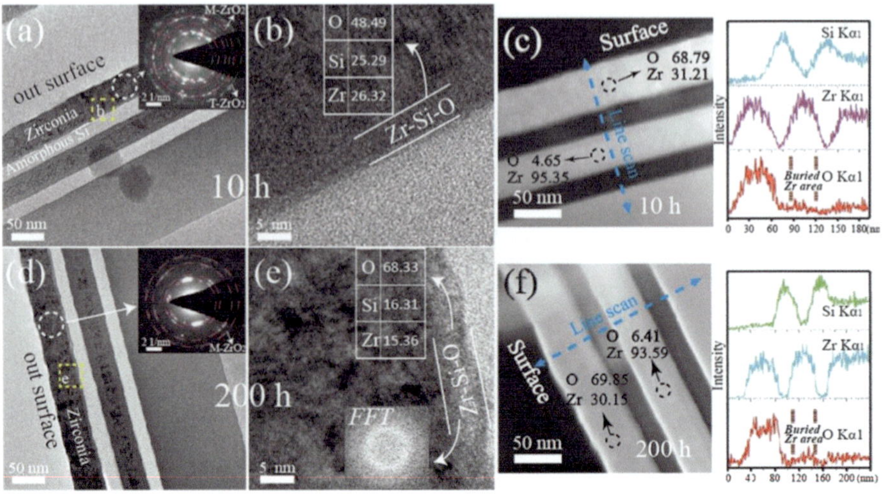

Fig. 5.4 **a–c** Cross-sectional morphology of the oxidized Zr/Si multilayers at 400 °C for 10 h. Figure 5.4b corresponds to the dotted rectangular box in Fig. 5.4a. **c** High-angle annular dark field (HAADF) images and EDS line-scanning profile of Zr/Si multilayers after 10 h. **d–f** Cross-sectional morphology of the oxidized Zr/Si multilayers at 400 °C for 200 h. Figure 5.4e corresponds to the dotted rectangular box in Fig. 5.4d. **f** HAADF images and EDS line-scanning profile of Zr/Si multilayers after 200 h

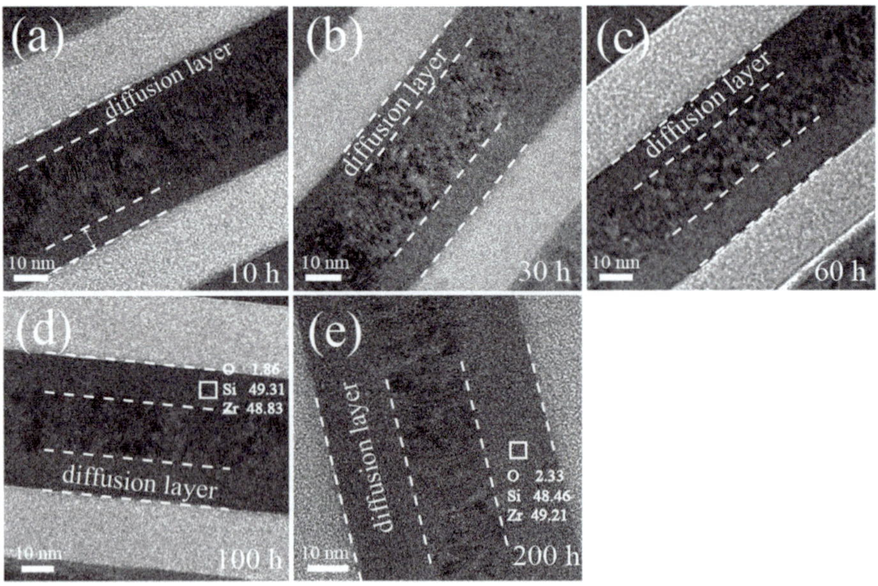

Fig. 5.5 Morphology and thickness of the Zr-Si amorphous layer as a function of oxidation time at 400°C

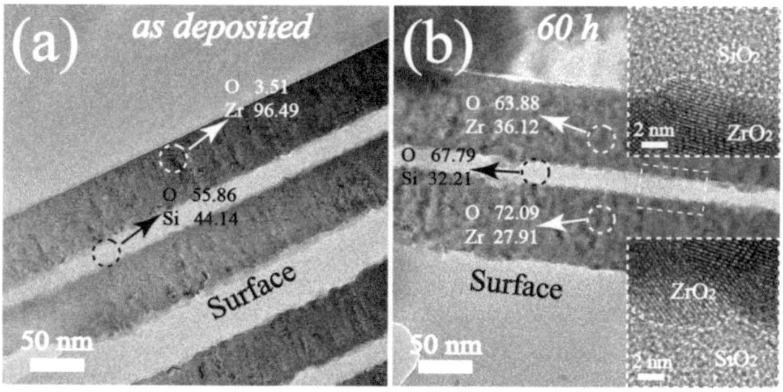

Fig. 5.6 Fabricated (**a**) and oxidized (**b**) Zr/SiO_x/Zr system at 400 °C for 60 h. The interface structure after oxidation was shown in the inset of Fig. 5.6b

It has been reported that α-SiO_2 cannot block elemental diffusion at 800 °C, such as metal cations and O ion. Hence, the internal oxidation occurred [30]. Wang et al. had analyzed the reactions of Zr thin films with SiO_2 substrates. They found that when the temperature is sufficiently high, SiO_x dissociated in the presence of Zr [38]. As Zr has a high heat of formation while reacting with oxygen to form ZrO_2, oxygen can release from SiO_x and diffuse into the whole region of Zr uniformly. Special attention was also paid to the interface structure after oxidation, as shown in Fig. 5.6b. The top-ZrO_2/SiO_x interface structure was distinct without solid-state reaction, indicating poor mutual dissolution of ZrO_2 and silica as no amorphous ZSO layer was formed between ZrO_2 and SiO_x.

Formation and Function of ZSO: From Fig. 5.4e, when the ZSO layer is formed, its thickness is almost constant, indicating high stability of the *ZrO_2/ZSO/Si* system. To find the effect of elemental diffusion on the phase formation sequence, two multi-layer systems were fabricated with a consistent Si layer (5 nm) and variable top Zr layer. As shown in Fig. 5.7a, the top Zr layer is approximately 50 nm before oxidation. After 60 h of oxidation, almost all the Si was consumed and transferred into the amorphous ZSO layer. The second Zr layer was unoxidized during this process. Therefore, a single ZSO layer can provide a diffusion barrier for oxygen flux regardless of the thickness of the Si layer. In Fig. 5.7b, the Zr thickness was reduced to 12 nm. The inset in Fig. 5.7b shows the TEM image of as-deposited Zr (12 nm)/Si (5 nm)/Zr (12 nm). After 60 h of oxidation, the surficial Zr transferred into ZrO_2 and the thickness increased by 40%. The sandwiched Si was still in element state, while the second Zr layer was incorporated by 22.56 at. % O. The enrichment of oxygen (Fig. 5.7b) proves the significance of ZSO in the oxygen invasion barrier. It has to be mentioned that the ZSO layer was not formed under this condition. When the thickness of the top Zr layer was reduced to 12 nm, the absence of ZSO was explained by two competing processes. The oxidation of the top Zr layer is a reaction-controlled process (I) [39]. On the other hand, due to the negative heat of mixing of the Zr–Si system (−84 kJ/mol), an intermixing occurs at the Zr/Si multilayer interface thermodynamically, which is a diffusion-controlled silicidation process (II), where the diffused silicon reacts with Zr across the interface. Figure 5.8 shows the growth kinetics of intermixing Zr–Si layer and the oxidation kinetics of Zr as a function of oxidation time at 400 °C. The growth kinetics of Zr–Si transition layer matches the logarithmic law, while the oxidation rate of zirconium obeys a linear law. The reaction-controlled process I has a higher reaction rate than that of the diffusion-controlled process II. When the Zr (top layer) is thick enough, there is sufficient time for the silicidation process (II) to occur. In such a case, $ZrSi_x$ is formed prior to the formation of ZSO. During the following diffusion of oxygen, the thin Zr–Si silicide was incorporated with O to form ZSO. When the thickness reduced to 12 nm (Fig. 5.7b), the oxidation process (I) finished instantly. With the formation of ZrO_2/Si bonding system, there will be no time for metallic Zr to react with Si on the interface because Zr was closely bonded in ZrO_2. Compared to Zr silicides, ZrO_2 is thermodynamically favored at medium temperature. The absence of ZSO will cause the oxidation of Zr in the second layer, as shown in Fig. 5.7b.

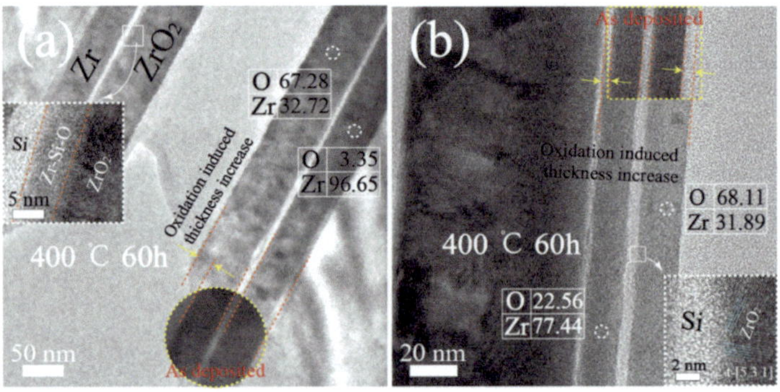

Fig. 5.7 Morphology of the two multilayer systems with a consistent Si layer (5 nm) and variable Zr layer after 60 h of oxidation. **a** 50 nm for Zr layer; **b** 12 nm for Zr layer

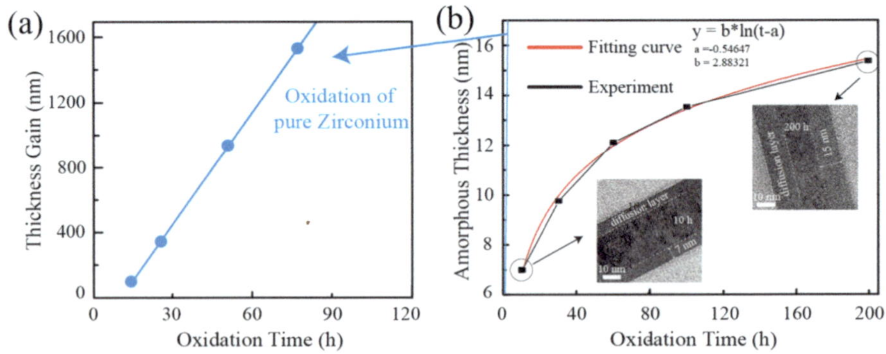

Fig. 5.8 **a** Oxidation kinetics of pure zirconium growth and **b** growth kinetics of the intermixing Zr–Si layer as a function of oxidation time at 400 °C

To examine the chemical stability between Si and ZrO_2, an asymmetric three-layer structure was fabricated using *Zr/Si/*ZrO_x, as shown in Fig. 5.9a. The structure transformed to ZrO_2/*Si*/ZrO_x after 2 h of oxidation at 400 °C, as observed from Fig. 5.9b. The inset EDS line-scan result demonstrates that silicon has not oxidized while the zirconium layers have oxidized on both sides of the Si layer. The ZrO_2/*Si*/ZrO_x structure was quite stable and remained unchanged even when the oxidation time was increased to 200 h, as observed from Fig. 5.10. The insets in Fig. 5.10 exhibit the interface morphology on both sides of the Si layer. It is interesting to find that at the interface of *Si*/ZrO_x, no elemental diffusion occurred on both sides and Si keeps a sharp and stable interface with Zr oxide.

Performance Under PWR Conditions: To further evaluate the oxidation behavior of Zr/Si multilayers in a subcritical environment (PWR), the corrosion experiment was performed at 320 °C and 15.5 MPa for 10–200 h. The deposited multilayer composed of twelve Zr/Si bilayers having a 25 nm Zr/20 nm Si layer thickness. From the cross-sectional morphology of the prepared Zr/Si multilayers shown in Fig. 5.11a, the multilayers exhibited a uniform structure and distinct interface at a large scale. Figure 5.11b, c shows the bright-field images and EDS line

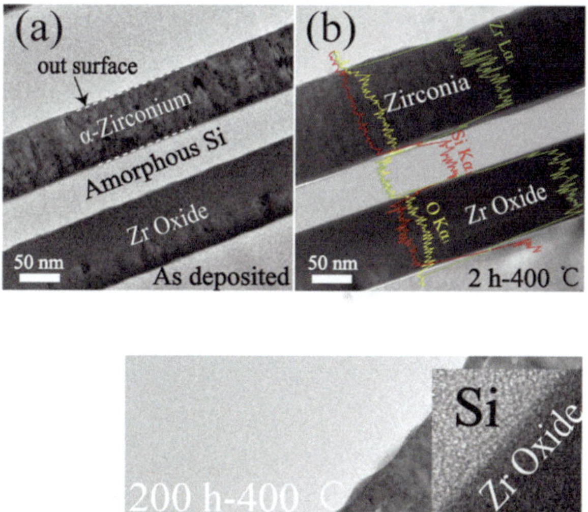

Fig. 5.9 Microstructure evolution and elemental distribution of the three-layered Zr/Si/ZrO_x coatings before and after 2 h oxidation at 400 °C

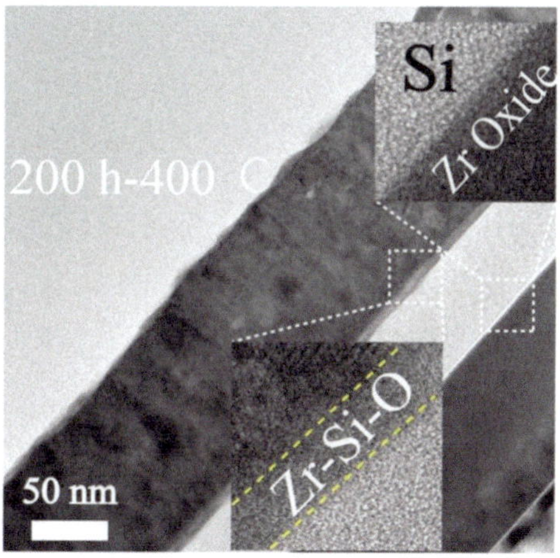

Fig. 5.10 Interface microstructure of the three-layered Zr/Si/ZrO_x coatings at 400 °C for 200 h

profile of the Zr/Si multilayers exposed to the corrosion environment at different times. The EDS data clearly demonstrate that the top Zr layer is oxidized while the remaining layers show little uptake of oxygen after a long time. It can be observed that the capping ZrO_2 layer is quite dense and uniform. Figure 5.11d shows a ZSO layer existing beneath the top oxide layer, similar to that in air oxidation. The top Zr layer would quickly oxidize under both, subcritical or air annealing conditions. The formation of the ZSO layer inhibits the dissolution of Si in water and provides a barrier for oxygen diffusion.

The Zr was designed as top layer in this work to act as a barrier to dissolution of Si, mainly for the application in nuclear reactor system. If Si becomes top layer, significant surface recession would occur [28–30]. Fournier and Rowe reported that the solubility of amorphous SiO_2 in pure water increased exponentially with increasing water temperature and found a linear relationship between the solubility and autoclave pressure. As a prototype research in short period (~200 h) oxidation, the protective effect can be achieved by the multilayer structure in this work. The anti-irradiation performance is expected in future.

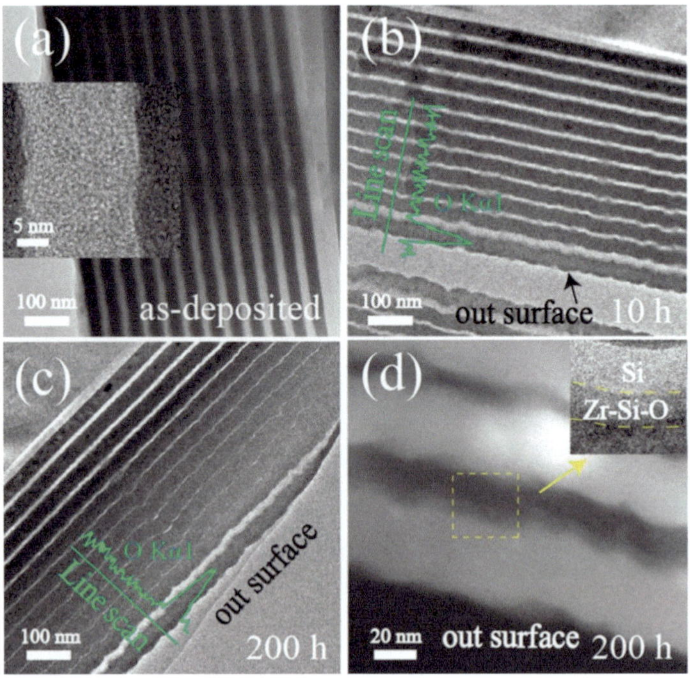

Fig. 5.11 **a–c** Cross-sectional morphology of multilayer composed of twelve Zr/Si bilayers having a 25 nm Zr/20 nm Si layer thickness at 320 °C and 15.5 MPa after 10 h-200 h of oxidation. **d** Formation of ZSO under the first ZrO_2 layer after 200 h of oxidation. The insets in **b**, **c** show the corresponding EDS line-scan profile of oxygen

5.4 Conclusion

Different multilayer structures based on Zr/Si system were deposited using the HiTUS system and oxidized in different conditions. The as-deposited multilayer structure consists of a nanocrystalline Zr layer (top) and an amorphous Si layer. The formation of Zr-Si intermixing layer as a bonding during oxidation process is a key point in this work, which improves the structural stability of the multilayer coating. No spallation or delamination occurred during the oxidation process. Moreover, an amorphous Zi–Si–O (ZSO) intermediate layer was formed in situ between top Zr and Si, which shows a mixture bonding characteristic of Zr–O and Si–O. The ZSO layer derived from the Zr-Si intermixing process prior to the oxygen incorporation is determined by a competition between silicidation and oxidation. The formation of ZSO contributes to the excellent oxidation resistance of $ZrO_2/ZSO/Si$. Additionally, the unique materials chemistry of amorphous Si synergistically contributes to the oxygen barrier of the multilayer systems. The Zr/Si multilayer structure also exhibits excellent oxidation resistance after an oxidation under PWR condition of 320 °C and 15.5 MPa.

References

1. Motta AT, Couet A, Comstock RJ (2015) Corrosion of zirconium alloys used for nuclear fuel cladding. Annu Rev Mater Res 45:311–343
2. Cao G, Ren Y, Wu K, Yao H, Hu J, Shao G, Yuan G (2019) Life extension strategy and research progress of zirconium alloy cladding applied to light water reactors. Surf Technol 48:69–81
3. Zeng B, Fan H, Chang H, Wang J (2019) Progress in preparation of zirconium alloy cladding surface coatings. Surf Technol 48:106–113
4. Cao G, Yang L, Yuan G, Hu J, Shao G, Yan L (2020) Chemical diversity of iron species and structure evolution during the oxidation of C14 Laves phase Zr $(Fe, Nb)_2$ in subcritical environment. Corros Sci 162:108218
5. Hu J, Yang L, Cao G, Yun Y, Yuan G, Yue Q, Shao G (2016) On the oxidation behavior of $(Zr, Nb)_2Fe$ under simulated nuclear reactor conditions. Corros Sci 112:718–723
6. Cao G, Yun Y, Yang L, Yuan G, Yue Q, Shao G, Hu J (2017) The formation and stacking faults of Fe and Cr containing Laves phase in Zircaloy-4 alloy. Mater Lett 191:203–205
7. Yuan G, Cao G, Yue Q, Yang L, Yun Y, Shao G, Hu J (2016) Formation and fine-structures of nano-precipitates in ZIRLO. J Alloy Compd 687:451–457
8. Han X, Wang Y, Peng S, Zhang H (2019) Oxidation behavior of FeCrAl coated Zry-4 under high temperature steam environment. Corros Sci 149:45–53
9. Ribis J, Wu A, Brachet JC, Barcelo F, Arnal B (2018) Atomic-scale interface structure of a Cr-coated Zircaloy-4 material. J Mater Sci 53:9879–9895
10. Maier B, Yeom H, Johnson G, Dabney T, Walters J, Xu P, Romero J, Shah H, Sridharan K (2019) Development of cold spray chromium coatings for improved accident tolerant zirconium-alloy cladding. J Nucl Mater 519:247–254
11. Zhang Z, Lai DMY, Lim SH, Chai J, Wang S, Jin H, Pan J (2018) Isothermal oxidation of the Ti_2AlC MAX phase coatings deposited by kerosene-fuelled HVOF spray. Corros Sci 138:266–274
12. Tang C, Steinbrueck M, Stueber M, Grosse M, Yu X, Ulrich S, Seifert HJ (2018) Deposition, characterization and high-temperature steam oxidation behavior of single-phase Ti_2AlC-coated Zircaloy-4. Corros Sci 135:87–98

13. Meng C, Yang L, Wu Y, Tan J, Dang W, He X, Ma X (2019) Study of the oxidation behavior of CrN coating on Zr alloy in air. J Nucl Mater 515:354–369
14. Yeom H, Maier B, Mariani R, Bai D, Fronek S, Xu P, Sridharan K (2017) Magnetron sputter deposition of zirconium-silicide coating for mitigating high temperature oxidation of zirconium-alloy. Surf Coat Tech 316:30–38
15. Cao S, Wang Q, Hu J, Fu Z, Bai K, Shao G, Cao G (2018) Dominant growth of higher manganese silicide film on Si substrate by introducing a Si oxide capping layer. J Alloy Compd 740:541–544
16. Gao Y, Yang T, Xue J, Yan S, Zhou S, Wang Y (2011) Radiation tolerance of Cu/W multilayered nanocomposites. J Nucl Mater 413:11–15
17. Zare A, Su Q, Gigax J, Shojaee SA, Nastasi M, Shao L, Lucca DA (2017) Effects of ion irradiation on structural and mechanical properties of crystalline Fe/amorphous SiOC nanolaminates. Acta Mater 140:10–19
18. Kim I, Jiao L, Khatkhatay F, Martin MS, Lee J, Shao L, Zhang X, Swadener JG, Wang YQ (2013) Size-dependent radiation tolerance in ion irradiated TiN/AlN nanolayer films. J Nucl Mater 441:47–53
19. Uglov VV, Abadias G, Zlotski SV, Kvasov NT, Saladukhin IA, Malashevih AA (2018) Blister formation in ZrN/SiN multilayers after He irradiation. Surf Coat Tech 344:170–176
20. Nogueira GM, Banerjee D, Cohen RE, Rubner MF (2011) Spray-layer-by-layer assembly can more rapidly produce optical-quality multistack heterostructures. Langmuir 27:7860–7867
21. Hattar K, Demkowicz MJ, Misra A, Robertson IM, Hoagland RG (2008) Arrest of He bubble growth in Cu-Nb multilayer nanocomposites. Scripta Mater 58:541–544
22. Manukyan KV, Tan W, deBoer RJ, Stech EJ, Aprahamian A, Wiescher M, Rouvimov S, Overdeep KR, Shuck CE, Weihs TP, Mukasyan AS (2015) Irradiation-enhanced reactivity of multilayer Al/Ni nanomaterials. Acs Appl Mater Inter 7:11272–11279
23. Lei Z, Liu Y, Ma F, Song Z, Li Y (2016) Oxidation resistance of TiAlN/ZrN multilayer coatings. Vacuum 127:22–29
24. Alat E, Motta AT, Comstock RJ, Partezana JM, Wolfe DE (2016) Multilayer (TiN, TiAlN) ceramic coatings for nuclear fuel cladding. J Nucl Mater 478:236–244
25. Zhou Z, Rainforth WM, Lewis DB, Creasy S, Forsyth JJ, Clegg F, Ehiasarian AP, Hovespian PEh, Münz WD (2004) Oxidation behaviour of nanoscale TiAlN/VN multilayer coatings. Surf Coat Tech 177:198–203
26. Mori T, Noborisaka M, Watanabe T, Suzuki T (2012) Oxidation resistance and hardness of TiAlSiN/CrAlYN multilayer films deposited by the arc ion plating method. Surf Coat Tech 213:216–220
27. Northwood DO (1985) The development and applications of zirconium alloys. Mater Design 6(2):58–70
28. Duan Z, Yang H, Satoh Y, Murakami K, Kano S, Zhao Z, Shen J, Abe H (2017) Current status of materials development of nuclear fuel cladding tubes for light water reactors. Nucl Eng Des 316:131–150
29. Danko VA, Indutnyi IZ, Lysenko VS, Maidanchuk IY, Minko VI, Nazarov AN, Tkachenko AS, Shepelyavyi PE (2005) Kinetics of structural and phase transformations in thin SiOx films in the course of a rapid thermal annealing. Semiconductors 39(10):1197–1203
30. Čapek J, Hřeben S, Zeman P, Vlček J, Čerstvý R, Houška J (2008) Effect of the gas mixture composition on high-temperature behavior of magnetron sputtered Si-B-C-N coatings. Surf Coat Tech 203(5–7):466–469
31. Park JH, Kim HG, Park JY, Jung Y, Park D, Koo Y (2015) High temperature steam-oxidation behavior of arc ion plated Cr coatings for accident tolerant fuel claddings. Surf Coat Tech 280:256–259
32. Zhang X, Li Y, He X, Sun Y, Pang S, Su G, Liu X, Yang Z (2015) Influence of Cr addition on microstructure and mechanical properties of Zr-based alloys corresponding to Zr–C–Cr system. J Alloy Compd 640:240–245
33. Zhou BX, Yao MY, Li ZK, Wang XM, Zhou J, Long CS, Liu Q, Luan BF (2012) Optimization of N18 Zirconium alloy for fuel cladding of water reactors. J Mater Sci Technol 28(7):606–613

34. Likhanskii VV, Evdokimov IA (2009) Effect of additives on the susceptibility of zirconium alloys to nodular corrosion. J Nucl Mater 392(3) 447–452
35. Mitton DB, Yoon JH, Cline JA, Kim HS, Eliaz N, Latanision RM (2000) Corrosion behavior of Nickel-Based alloys in supercritical water oxidation systems. Ind Eng Chem Res 39(12):4689–4696
36. Deal BE (1963) The oxidation of silicon in dry oxygen, wet oxygen, and steam. J Electrochem Soc 110:527–533
37. Chou K, Chu PW, Marquis EA (2018) Early oxidation behavior of Si-coated titanium. Corros Sci 140:297–306
38. Takeuchi A, Inoue A (2005) Classification of bulk metallic glasses by atomic size difference, heat of mixing and period of constituent elements and its application to characterization of the main alloying element. Mater Trans 46:2817–2829
39. Yeom H, Lockhart C, Mariani R (2018) Evaluation of steam corrosion and water quenching behavior of zirconium-silicide coated LWR fuel claddings. J Nucl Mater 499:256–267

Chapter 6
"Irradiation–Oxidation" Correlation and Phase Stabilization Mechanism of Zirconia Coating

Abstract ZrO_2 phase formed in conventional oxidation environment has poor stability and cannot effectively protect Zr alloys. In this work, ZrO_2 coating with special structure was prepared on Zr alloy, and the defects' structure was introduced by Ar^{8+} bombardment. Martensitic phase transitions were found in ZrO_2 coatings, which were caused by lattice distortion and gradient mismatch at the interface. In addition, Ar^{8+} bombardment induces the formation of interstitial-vacancy pairs in ZrO_2. The DFT model proves that this irradiated vacancy has an advantage in stabilizing T-phase ZrO_2 compared to natural oxygen vacancy. A corrosion resistance mechanism based on unique interface and defect state was proposed.

Keywords Zirconium · Oxide coatings · Oxidation · TEM · Raman spectroscopy

6.1 Introduction

Zirconium alloy cladding materials play a key role in the development of advanced nuclear reactors, whose main function is to effectively isolate the fuel from the coolant in the reactor core. However, oxidation corrosion of zirconium alloy in coolant environment is one of the main reasons for its failure. The corrosion process of zirconium alloys can be regarded as an electrochemical reaction in which different diffused substances migrate through the oxide film driven by an electrochemical gradient. In order to improve the stability of the fuel cladding, researchers have carried out a lot of work to reduce the harmful reaction between zirconium alloys and high-temperature water [1–5].

Some progress has been made in the study of adding alloying elements to zirconium alloys, but it should be noted that substantial adjustment of alloy composition to improve its stability may lead to a decline in other key properties [6, 7]. In addition, coating technology is also used to enhance the corrosion resistance of zirconium alloys. Depending on the preparation method, the coating thickness is usually between a few microns to tens of microns. Recently, J.C. Brachet et al. developed a chromium (Cr) coating by physical vapor deposition (PVD) technology

J. Hu and G. Cao, *Zirconium Alloy Coatings*,
https://doi.org/10.1007/978-981-96-5028-6_6

and investigated in depth the atomic interface between the zirconium substrate and the chromium coating [8, 9]. In addition, studies have shown that materials such as MAX phases (such as Ti_3SiC_2), CrN, and CrC can be used as substitutes for zirconium alloy coatings [10, 11]. However, due to differences in physical properties and lattice interfaces, these coatings tend to crack rapidly.

Under the normal operation conditions of the reactor, an oxide film will naturally form on the surface of the zirconium alloy, which plays a passivation protection role. Some researchers have attempted to prepare zirconia coatings on the surface of zirconium alloys by different techniques, such as high-pressure steam treatment or high-temperature heating in dry air/oxygen [12, 13]. In addition, surface modification by microarc oxidation or plasma electrolytic oxidation technology can significantly improve the corrosion resistance of zirconium alloys [14, 15]. However, oxide coatings usually have poor toughness, high hardness, and defects such as tiny holes and wavy cracks, which provide diffusion channels for corrosive media, and local water radiolysis products may also be absorbed by the pores in the oxide. In addition, high-temperature treatment will inevitably cause certain damage to the zirconium alloy matrix.

The effect of ion irradiation on oxide layer is also concerned. Many researchers evaluate the radiation resistance of materials through ion bombardment experiments, focusing on phase transitions and structural changes [16, 17]. In addition, studies have shown that Ar ion irradiation can regulate the defect state and properties in oxides [18]. However, there are relatively few studies on the use of ion irradiation to improve the corrosion resistance of materials in nuclear environment. Nevertheless, reducing the corrosion rate by controlling the microstructure evolution of the oxide coating on the alloy surface remains a technical challenge.

Although the oxide coating with good corrosion resistance can effectively inhibit the oxidation of zirconium alloys, the oxide coating grown in situ usually has large pores and stresses, making it easy to crack under working conditions. In this study, we used high target utilization sputtering (HiTUS) technology for coating treatment, which can achieve high ion density sputtering at low temperatures. At the same time, Ar bombardment is not only used as a working environmental condition during coating deposition, but also as a means of material modification. Argon was chosen because of its slow physical properties and significant radiation damage effects. In addition, the mechanism of corrosion resistance of irradiated modified ZrO_2 coating in subcritical water environment was studied. This work is helpful to further reveal the synergistic effect of oxidation process and irradiation effect on coating phase transition and defect structure evolution.

6.2 Results

Controllable oxygen defects in ZrO_2 coatings were prepared by Far-source plasma plasma sputtering control. In order to investigate the effect of different parameters in the far-source plasma magnetron sputtering system on the microstructure of ZrO_2

coatings, we set the plasma excitation power PLS, target bias power DC and oxygen partial pressure as a single variable, respectively, uniformly adjusted the deposition time of the films to one hour, and then characterized the microstructures of the films in an attempt to find out the correlation between the different parameters and the quality of the films. Seven sets of samples were prepared using the parameters shown in Table 6.1.

The thickness of the coatings of seven groups of samples after sputtering for 1 h was measured by a step meter, and the film deposition rates deposition rates (DR, nm/h) are recorded in Table 6.1, from which it can be seen that both PLS and DC have a significant effect on the film deposition rate in this experiment, whereas the partial pressure of oxygen has almost no effect on it. Varying the plasma excitation power (PLS) from 600 to 1000 W increased the film deposition rate from 42 nm/h to 57 nm/h, and a similar phenomenon can be seen in varying the target bias power (DC). Thus, an increase in film deposition rate can be achieved by increasing either PLS or DC alone. Increasing PLS increases the plasma density and thus the number of plasmas impinging on target atoms, which increases the number of target atoms involved in the reaction and increases the deposition rate, while increasing DC increases the number of target atoms and reactive gas atoms involved in the reaction, which increases the reaction rate, i.e., increases the deposition rate.

Figure 6.1 shows the XRD spectra of the thin-film samples under different preparation parameters. Among them, the black, blue, and red curves in Fig. 6.1a correspond to the XRD spectra of the film samples with different PLS energy values for samples 1#, 2#, and 3#, respectively, i.e., when the DC power was fixed at 250 W and the oxygen flow rate was fixed at 7 sccm. In Fig. 6.1, the diffraction peaks with diffraction angles located at $2\theta = 24.0°$, 28.2°, 31.5°, 34.2°, 40.7°, 49.3°, 54.1°, 55.3°, and 59.8° correspond to the PDF standard card (PDF#37–1484) of M-ZrO_2, respectively, where 28.2°, 31.5°, and 34.2° are the PDF card's three strong peaks, which correspond to the (−111), (111), and (002) crystal planes of M-ZrO_2, respectively. The 30.3° and 50.4° diffraction peaks in the Figure also correspond to the (111) and (220) crystal planes of the PDF standard card (PDF#17–0923) for T–ZrO_2, respectively. As can be seen in Fig. 6.1, we successfully prepared zirconia thin films with three different parameters consisting of monoclinic (M-ZrO_2) and tetragonal (T–ZrO_2) phases, but the contents of the two phases differ and the difference mainly lies in the change of the strength of the diffraction peaks of the zirconia in the 30.3° tetragonal phase within the green box in the figure. When the PLS is 600 W, the thin-film material is mainly composed of monoclinic phase zirconia with trace amounts of tetragonal

Table 6.1 Preparation parameters and corresponding deposition rates

Number	1#	2#	3#	4#	5#	6#	7#
PLS (W)	600	800	1000	800	800	800	800
DC (W)	250	250	250	200	300	250	250
O_2 (sccm)	7	7	7	7	7	6	8
DR (nm/h)	42	52	57	46	53	49	51

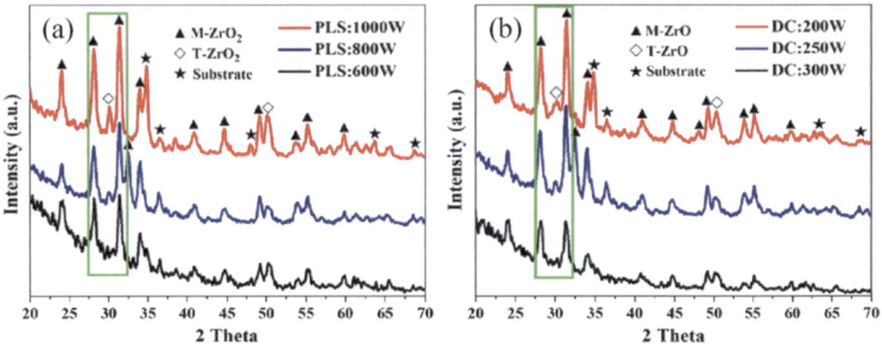

Fig. 6.1 XRD spectra of coatings prepared with different parameters

phase zirconia. As the PLS gradually increases from 600 to 800 W and then to 1000 W, we can see that the tetragonal phase zirconia diffraction peaks in the green frame are also increasing, and the content of tetragonal phase in the film is increasing. This is due to the fact that when the PLS is higher, the HiTUS system accordingly excites a higher density of plasma per unit gas flow rate, resulting in an increase in the number of plasma impinging on the target atoms, which increases the number of target atoms participating in the reaction, and tends to form more nuclei on the substrate, resulting in finer grains of the material, and these finer zirconia grains promote the formation of tetragonal phase due to the size effect.

The black, blue, and red curves in Fig. 6.1b correspond to the XRD spectra of samples 4#, 2#, and 5#, i.e., the thin-film samples corresponding to different target bias DCs when PLS is fixed at 800 W and oxygen flow rate is fixed at 7 sccm. The zirconia under these three different parameters is also all composed of monoclinic and tetragonal phases, and there is also a difference between the two phases mainly in the change of the strength of the zirconia diffraction peaks of the 30.3° tetragonal phase in the green box in the figure. When the DC is 300 W, the thin-film material is mainly composed of monoclinic phase zirconia and trace amounts of tetragonal phase zirconia. The DC component in the HiTUS system provides the accelerating voltage for the plasma entering the chamber, and when the DC is low (but the sputtering threshold of the target itself must be exceeded), the low target bias reduces the probability of collisional binding of plasma and target atoms per unit density, which results in more defects in the material grown on the substrate. The change in lattice constants caused by defects increases the fractional distortion, which acts as a resistance to the phase transition of zirconia and stabilizes the tetragonal phase of zirconia that exists in the early stages of nucleation. From the above XRD results, we can see that either increasing the plasma excitation power (PLS) alone or decreasing the DC target power alone can lead to more tetragonal zirconia in the prepared thin-film materials. That is, we can design the composition of the thin-film material by regulating the size of PLS/DC in the HiTUS system, and the content of tetragonal phase increases with the increase of PLS/DC.

The reaction atmosphere oxygen is introduced into the HiTUS system as a source of oxygen in ZrO_2 coatings, which has an important effect on the microstructure of the coating materials. Oxygen enters the chamber at a small flow rate, which can form substoichiometric ZrO_2 with oxygen vacancies. In order to discuss the effect of oxygen partial pressure on the coating material, we further did XPS, Raman characterization for samples 6#, 2#, 7# (PLS:800W, DC:250W, oxygen flow rate: 6, 7, 8 sccm). Figure 6.2a shows the Raman spectra of ZrO_2 films at different oxygen flow rates (6, 7, 8sccm). Monoclinic and tetragonal phases of zirconia were found in each group of coatings. M-ZrO_2 belongs to the space group C52h (P21/c), and there are 36 lattice vibrational modes of M-ZrO_2: Gmono = 9 Ag + 9 Au + 9Bg + 9 Bu, of which 9 Ag and 9 Bg are Raman-active vibrational modes, 8 Au and 7 Bu are infrared-active vibrational modes, and there are remaining 1 Au and 2 Bu which are other acoustic vibrational modes. The Ag of M-ZrO_2 is at: 176, 187, 300, 475, 558, 635, and 760 cm^{-1}, and the Bg is at: 220, 333, 344, 380, 510, 536, and 613 cm^{-1}. T–ZrO_2 belongs to the space group C154h (P42/nmc), and T–ZrO_2 has six Raman-active vibrational modes (A1g + 2B1g + 3Eg), of which B1g: 147 cm^{-1}, Eg: 269 cm^{-1}, B1g: 315 cm^{-1}, Eg: 459 cm^{-1}, A1g: 600 cm^{-1} (very weak), and Eg: 643 cm^{-1}. Typically, we use the characteristic peaks of B1g (147 cm^{-1}) and Eg (260 cm^{-1}) to determine whether the tetragonal phase is present. We zoomed in on the B1g: 147 cm^{-1} region of tetragonal phase zirconia, and it is clear from the figure that the coatings prepared at different oxygen flow parameters are all composed mainly of monoclinic phases, and the constituent phases of the coatings are almost free of tetragonal phases when the oxygen flow rate is 8 sccm. When the oxygen flow rate is 7 and 6 sccm, the coating is a two-phase mixture of monoclinic and tetragonal phases, and the intensity of the B1g and Eg Raman peaks of T–ZrO_2 increases significantly with the decrease of the oxygen flow rate, and the experimental results match well with the above XRD data. This further supports that oxygen vacancies can help to form the tetragonal phase through the relaxation generated by the internal lattice distortion, and the phenomenon of stabilizing the tetragonal phase by doping Y^{3+}, Ca^{2+}, Mg^{2+}, Fe^{3+}, Er^{3+}, and Tb^{3+} was also introduced to introduce the mechanism of oxygen vacancies. Higher concentration of oxygen vacancies leads to lower T-phase formation energy. Tetragonal zirconia is of interest due to its low volume change during heating–cooling cycles. The XPS results in Fig. 6.2b show that the oxygen vacancies become apparent as the flow rate decreases. The stability of the martensitic phase transition and phase transformation of zirconia can have a significant impact on the coating, affecting the corrosion rate and oxidation process of the cladding tubes it protects. Although deposited T–ZrO_2 with natural oxygen vacancies is unstable during oxidation, prefabricated vacancy defects may somehow improve the stability of the coating during irradiation. Therefore, we chose a biphasic coating with oxygen vacancies (O:6) to maximize the content of tetragonal phases and thus investigated the structural evolution process of this biphasic structure of zirconia during in-heap oxidation and irradiation. Figure 6.2c, d shows the transmission electron microscopy bright-field image (BF) and selected area electron diffraction (SAED) patterns of the biphasic coating. In Fig. 3.2c, it can be clearly seen that the coating is homogeneous and dense (thickness of about 250 nm) without holes and cracks. In Fig. 5.2d, the coexistence of monoclinic and tetragonal phases is consistent with the Raman results in Fig. 5.2a.

Structural Evolution and Defect Types of Coatings Under Ion Irradiation. The zirconia coating was irradiated with Ar^{8+} heavy ions excited at a high pressure of 270 kv at a reference pressure of 5×10^{-5} Pa and room temperature to investigate the microstructural evolution of the film under irradiation damage in the reactor. The SRIM software was used to simulate the energy loss of ions in the target and the resulting defect distribution, in which Ar^{8+} heavy ions of a certain energy were struck into the designed ZrO_2 to obtain the simulated information of the 2D trace diameter as shown in figure. We used two irradiated ion doses of 4.32×10^{15} ions/cm^2 and 8.64×10^{15} ions/cm^2 for the samples. With the irradiation time and the

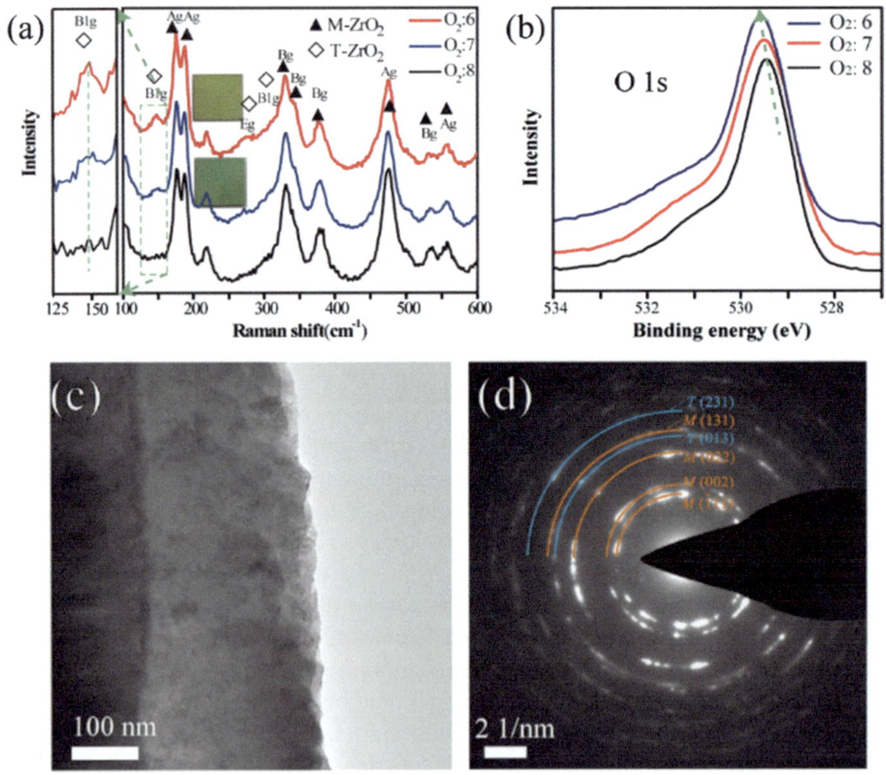

Fig. 6.2 **a** Raman spectra of ZrO_2 films with oxygen flow rates of 6, 7, and 8 sccm, respectively. **b** O 1s spectra for oxygen flow rates of 6, 7, and 8 sccm. **c** Bright-field (BF) TEM images of cross-sectional ZrO_2 film (6 sccm). **d** Selected-area electron diffraction (SAED) map of biphasic oxide (6 sccm)

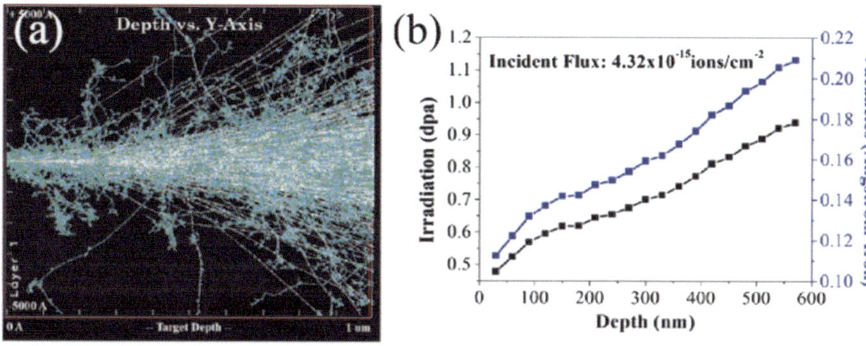

Fig. 6.3 **a** Simulated radiation damage of Ar ions. **b** Calculated radiation damage (DPA) and vacancy distribution at a dose of 4.32 × 1015 ions/cm^{-2}

number of delocalized atoms simulated by SRIM, we can obtain different values of irradiation damage for the coating material (thickness of 300 nm) corresponding to different numbers of irradiated ion doses. The variation of the number of delocalized atoms and irradiation damage values with depth due to incident ions at an irradiated ion dose number of 4.32×10^{15} ions/cm^2 is shown in Fig. 5.3. With the increase of irradiation depth, we can get the number of delocalized atoms produced by an incident ion per Å of travel through the blue curve in the Figure, the deeper the depth of ion irradiation in the range of 0–300 nm, the more vacancy defects are produced, and the number of delocalized atoms produced by an incident ion per Å of travel at 300 nm is about 0.15. The further calculated irradiation damage value DPA is shown as the black curve in figure. It can be seen that the irradiation damage value and the number of delocalized atoms follow the same basic trend in the thickness range of the coating, and the average irradiation damage value is about 0.6 dpa in the thickness range of the film, and since the irradiation damage value is linearly related to the irradiation ion dose number, the corresponding irradiation damage value for an irradiation ion dose number of 8.64×10^{15} ions/cm^2 is about 1.2 dpa. In the later part, we uniformly use the irradiation damage value of about 0.15 atoms per Å traveled. In the following analysis, we uniformly adopt the irradiation damage value DPA to quantitatively express the irradiation effect on the coating material.

Figure 6.4 shows the cross-section and surface morphology of the coating before and after irradiation. The thickness of the original coating is about 250 nm and consists of nanoparticles around 20 nm. Under irradiation conditions, no blistering or flaking was observed even at the highest irradiation dose (8.64×10^{15} ion / cm^{-2}, d inset) and the grain growth was negligible. The key strategy for designing high-performance radiation-resistant materials is based on the introduction of uniform, dense nanocrystalline particles, which provide large surface and grain-boundary active sites that can serve as effective absorption sites for irradiation defects. Thus, the nanocrystalline coatings prepared in the project are stable under irradiation, and the ideal grain size of less than 40 nm critical size will contribute to the stability of the coatings.

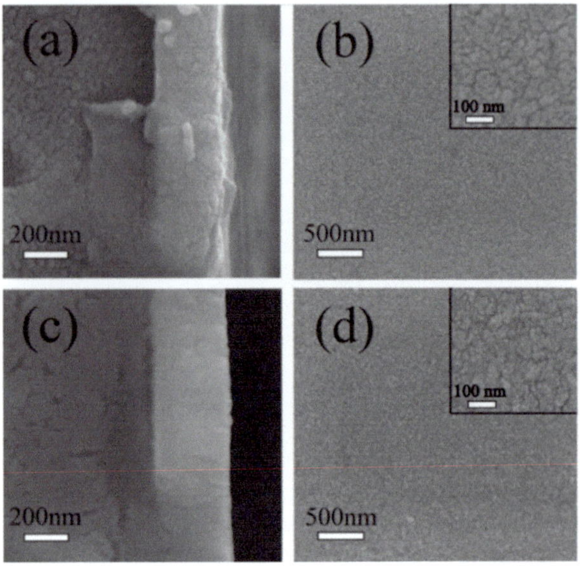

Fig. 6.4 SEM morphology of cross-section (**a**, **c**) and surface (**b**, **d**) of ZrO_2 films before (**a**, **b**) and after (**c**, **d**) irradiation at 1.2 dpa

Figure 6.5 shows the XRD spectra of the coating before and after Ar^{8+} irradiation. A clear phase transition from monoclinic to tetragonal zirconia is observed. As the irradiation dose increases, the dominance of M-ZrO_2 fades away and a T–ZrO_2 phase is formed with selective crystal planes of (111) and (200) (Fig. 6.5b). Normally, the critical DPA for the phase transformation of tetragonal zirconia should be higher than 1.0, while the minimum DPA for the phase transformation to occur in this experiment is 0.6. It suggests that the pre-existing oxygen vacancies in the zirconia may promote the phase transformation of monoclinic to tetragonal zirconia.

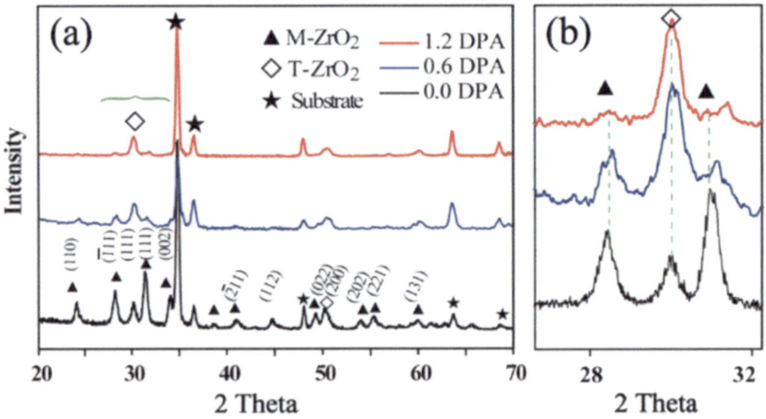

Fig. 6.5 **a** X-ray diffraction spectral evolution of ZrO_2 thin films before and after irradiation with 0.6, 1.2 DPA. **b** From the bracketed region in (**a**), showing XRD results for selected ranges in detail

The evolution of chemical states in the coating was studied by XPS in Fig. 6.6. As the irradiation dose increases, O 1s significantly shifts toward higher binding energies. This is attributed to Ar^{8+} "kicking out" oxygen from lattice sites and forming oxygen vacancies and interstitial oxygen. Figure 6.6c shows the results of O 1s peak splitting, corresponding to lattice oxygen, interstitial oxygen, and surface absorbed oxygen stabilizing around the binding energies of 529.4, 531.2, and 533.4 eV, respectively. As radiation damage increases, the peak value of interstitial oxygen becomes larger and larger. Radiation can provide enough energy for atoms to move from their original position to the interstitial position. Some interstitial oxygen may diffuse to the outer surface. As a result, the surface oxygen (533.4 eV, yellow) also increased. It should be pointed out that the Zr 3d peak shifts toward higher energy after irradiation, corresponding to higher valence states. Usually, the presence of oxygen vacancies leads to the presence of low valence Zr ions to neutralize the charge of ZrO_2, corresponding to a corresponding decrease in binding energy. However, in this study, irradiation may cause local lattice distortion, resulting in changes in the electron density around the atoms. The loss of outer electrons causes the inner electrons to bind more tightly to the atomic nucleus. It can be said that due to the irradiation of charged ions (Ar^{8+}), the electron density around Zr changes, causing the Zr 3d peak to shift toward higher energy positions.

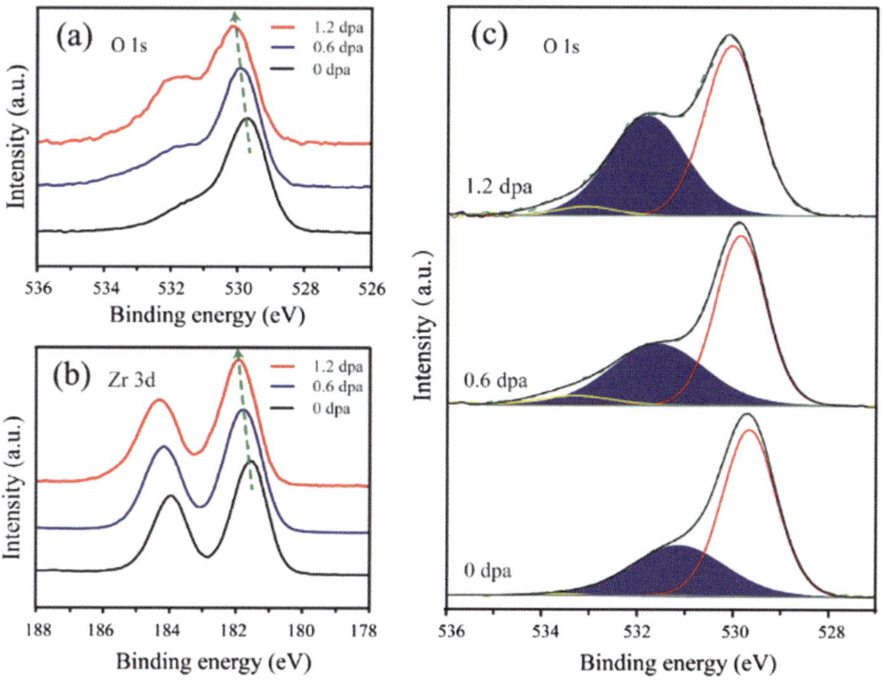

Fig. 6.6 XPS spectra of O 1s (**a**) and Zr 3d (**b**) in ZrO_2 thin films at different irradiation damages (0, 0.6, 1.2 DPA). The split-peak results of the oxygen gap are highlighted in blue in (**c**)

To confirm the evolution of the binding state, nanoindentation experiments were conducted with a maximum indentation depth maintained at 200 nm. In Fig. 6.7, the modulus significantly increases with the increase of irradiation dose. The illustration in Fig. 6.7 shows the average value of the elastic modulus. Elastic modulus is an inherent material property that is closely related to the binding energy between atoms. In addition, considering the high bonding ionicity of ZrO_2, the ionic bonding energy is actually the electrostatic force between ions: $f = KQ_1Q_2/r^2$, Q is the charge state of the ion, and r is the distance between two atoms. Therefore, the strength of ionic bonds depends on the valence state of the ions. The above XPS results indicate that Zr ions are in a higher bond energy state after ion irradiation. Based on the above discussion, it can be inferred that irradiation-induced Zr-O bond energy enhancement occurs.

We also used Raman spectroscopy to study the changes in physical phases before and after irradiation, as well as the bonding of the coating. The peak position at 149 cm^{-1} corresponding to the bending vibrations of O–Zr–O and Zr–O–Zr in the z-direction of tetragonal zirconia in Fig. 6.8 shifts toward higher wave numbers. This indicates that the newly formed T–ZrO_2 also has defects. The defect structure in the material strongly affects Raman spectra by generating specific peak shifts. In Raman spectroscopy, the shift of peaks toward higher wavenumbers is usually associated with shorter bond lengths. Based on the above XPS and nanoindentation tests, it can be said that the abnormal shrinkage of the average Zr–O bond caused by radiation. The ab initio calculations in the T–ZrO_2 lattice indicate that due to the contraction of

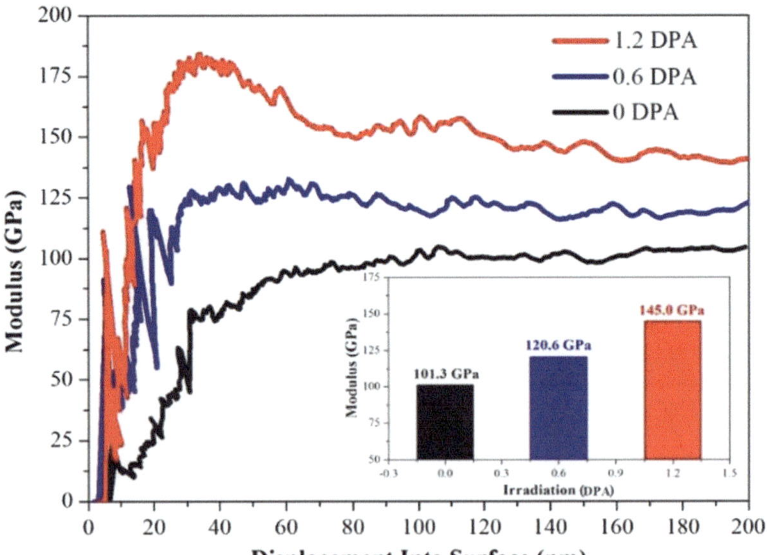

Fig. 6.7 Results of nanoindentation (0–200 nm depth) of ZrO_2 thin films at different irradiation damages (0, 0.6, 1.2 DPA). The inset shows the average value of elastic modulus in the range of 100–200 nm

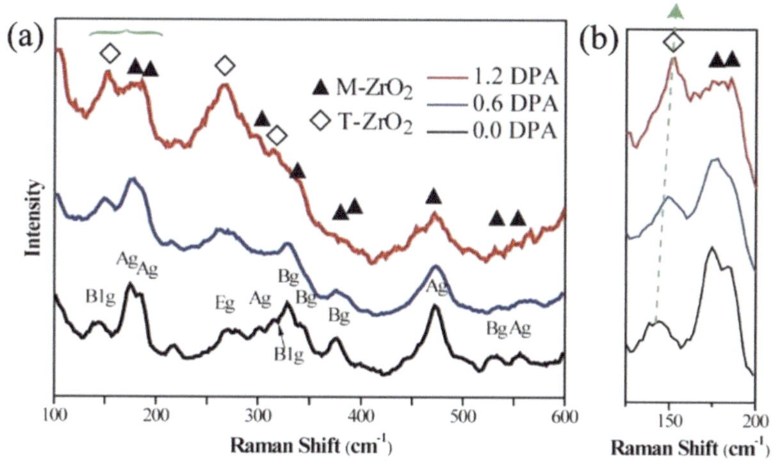

Fig. 6.8 **a** Raman spectra of ZrO_2 thin films under irradiation damage at 0, 0.6, and 1.2 DPA. **b** Raman spectra of selected T–ZrO_2 phases containing the frequency range of 125–200 cm^{-1}

interatomic distance, the activation energy for diffusion increases, and the diffusion coefficient tends to decrease. Therefore, after Ar^{8+} irradiation, oxygen diffusion will become difficult.

The TEM image of the cross-section ZrO_2 coating in Fig. 6.9 (the sample was subjected to 8.64×10^{15} ions/cm^{-2}) shows the growth of nanocrystalline T–ZrO_2 in M-ZrO_2 grains. The SAED plot in Fig. 6.9b shows the phase transition from monoclinic to tetragonal zirconia (compared to Fig. 5.2d). And it is found that there is an orientation relationship (OR) between monoclinic and tetragonal zirconia (Fig. 6.9c): M $[\bar{1}12]$//T [111] and M $(1\bar{1}1)$//T $(10\bar{1})$ and T [111]// T [001] [001] and T (110)/ / T $(11\bar{2})$. In Fig. 6.9d, we can see selected lattice distortion area (LDA), which is separated from the ideal lattice region by dashed lines. As shown, the atomic arrangement in the LDA approximates M-ZrO_2, with some of the atoms deviating from their normal lattice positions, and the inset in Fig. 6.9d shows atomic images similar to the M-phase. It should be noted that the T–ZrO_2 phase exists near the LDA, where disorder may propagate along the sublattice on the nanoscale during irradiation, and the M- to T-phase transition may be triggered by a gradient mismatch due to lattice distortion.

Anomalous growth of T–ZrO_2 is observed in Fig. 6.10. We propose a mechanism based on irradiation-induced lattice distortion and interface evolution. Figure 6.10b illustrates the interfacial region of irradiated M and nascent T–ZrO_2. Figure 6.10 (b_1, b_2) illustrates a magnified image of the selected region in Fig. 6.10b. The lattices close to M and T are indicated by the brown and blue dashed lines in Fig. 6.10 (b_1), respectively. The lattice arrangement clearly demonstrates an M-phase and T-phase inserted arrangement structure. As shown in Fig. 6.10(b_2), the M (101) plane is gradually rotated in the M-phase region. The selected M lattice cell (shown by the

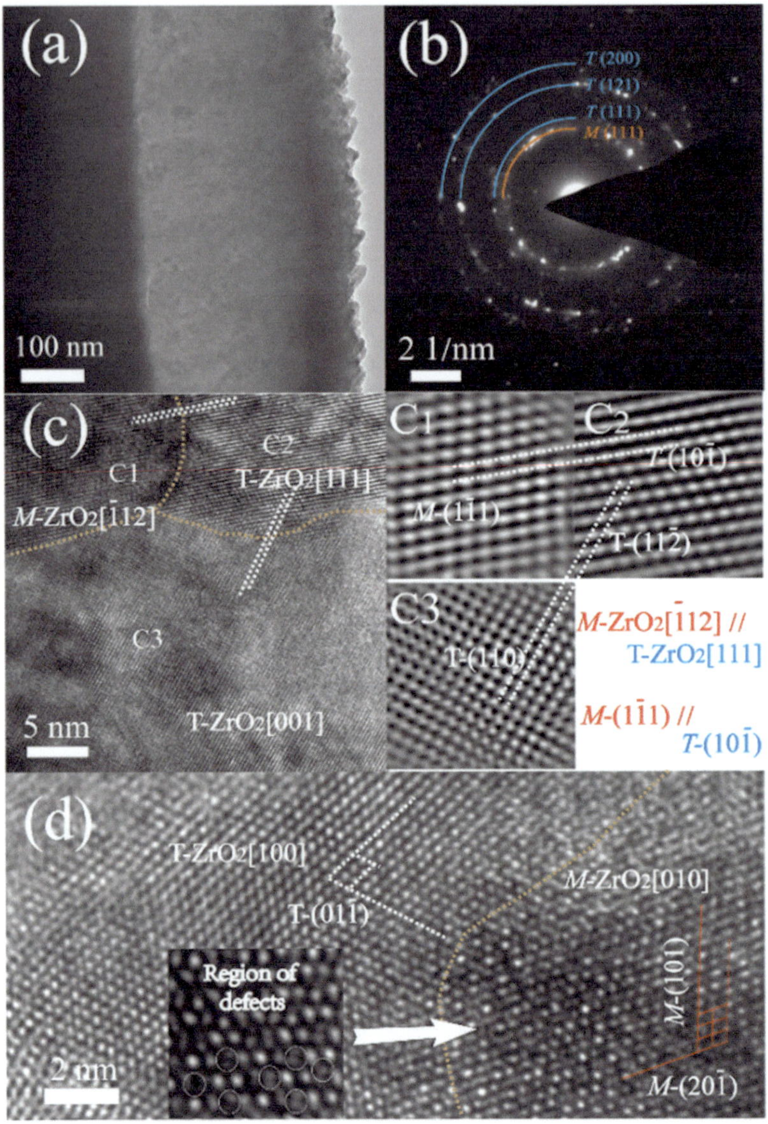

Fig. 6.9 Detailed structural characterization of ZrO_2 films at 8.64 × 10^{15} ion/ cm_{-2} irradiation dose: **a** bright-field TEM image of cross-sectional ZrO_2 film. **b** SAED structure of ZrO_2 coating after irradiation. **c** HRTEM images of monoclinic and tetragonal crystalline zirconia grains with polycrystalline boundaries. **d** HRTEM images of distorted regions

arrow) is defined as the starting point. The deformation of the six selected lattices is shown in Fig. 6.10(b_2). The rotation angle of M (101) is indicated by θ between M (101) and T (101), which changes from 0° to 19.9°. At the same time, the lattice is compressed by 17.14% along the normal direction of M (101), and the lattice is shifted to a T structure. The transformation model of the phase transition is drawn in Fig. 6.11. The stability of T–ZrO_2 can be enhanced by interfacial anchoring.

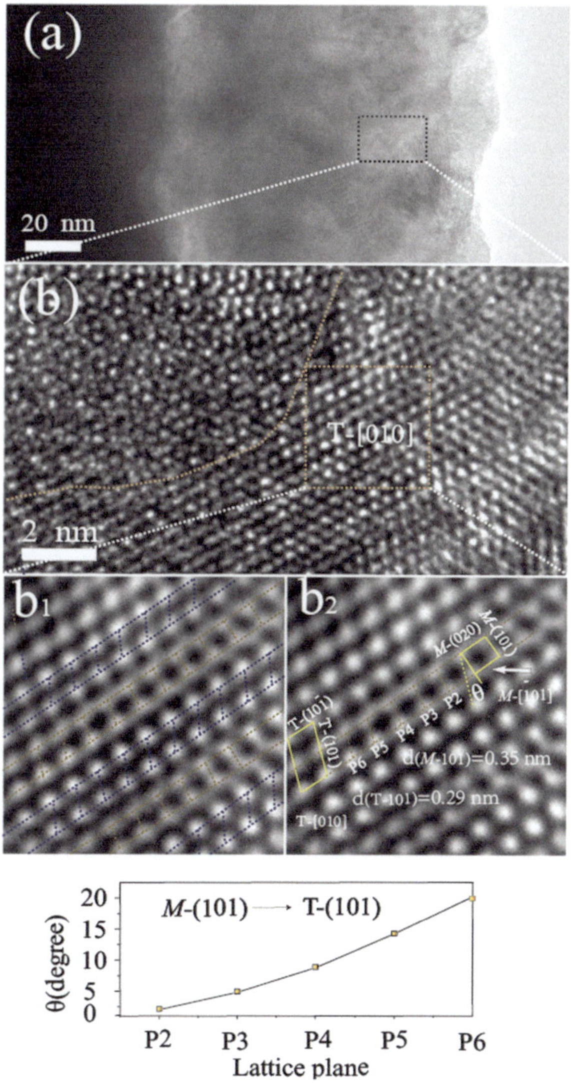

Fig. 6.10 **a** Bright-field TEM image of cross-sectional ZrO_2 membrane. **b** LDA and adjacent T-phase region. (b_1, b_2) Enlarged region of the dashed box, which shows the T–M gradient mismatch structure at the interface

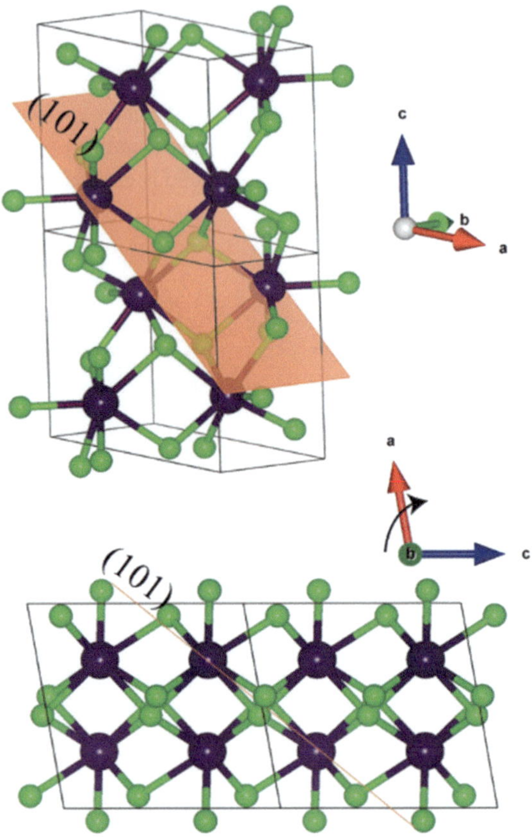

Fig. 6.11 3D structure showing the rotation of M (101) leading to the generation of T-phase ZrO_2

Thermodynamic stability is crucial for the oxidation resistance of coatings at high temperatures. The project performed theoretical calculations using VASP. Figure 6.12 shows the formation energy of zirconia, indicating that O vacancies favor the stabilization of the T-phase. When O vacancies reach a level of 10 at. %, T is more stable compared to M-phase. O vacancies can induce internal stresses around them, and the stresses are the origin of the stabilized T-phase. As shown in Fig. 5.6, more vacancy-gap pairs were formed during Ar^{8+} irradiation. In this work, the presence of interstitial oxygen along with oxygen vacancies after Ar^{8+} irradiation prompted further discussion of the role of vacancies-interstitials on T-phase stability. Based on the T-phase ZrO_2 with 10% O vacancies, different amounts of interstitial oxygen (2, 4, 6 and 8) were further introduced. As can be seen in Fig. 6.13, the energy of the system further decreases with the formation of interstitial O. The interstitial oxygen further increases the internal stresses and stabilizes the tetragonal phase when vacancy-interstitial pairs are formed on the lattice.

As an effective chemical conditioning method, the deposited ZrO_2 can be modified by Ar^{8+} irradiation. The project evaluated the thermodynamic stability of the

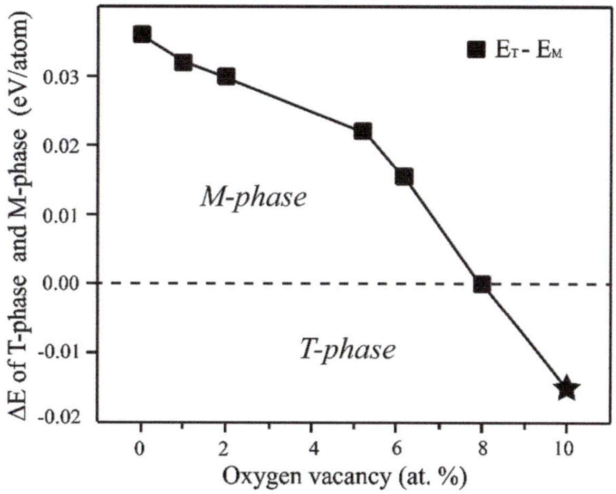

Fig. 6.12 Calculated formation energies of zirconia after full relaxation. Difference in the formation energies (ΔE) of the T-phase and the M-phase for O vacancies of 2% to 10%. It is proved that the O vacancy is responsible for the stabilization of the T-phase

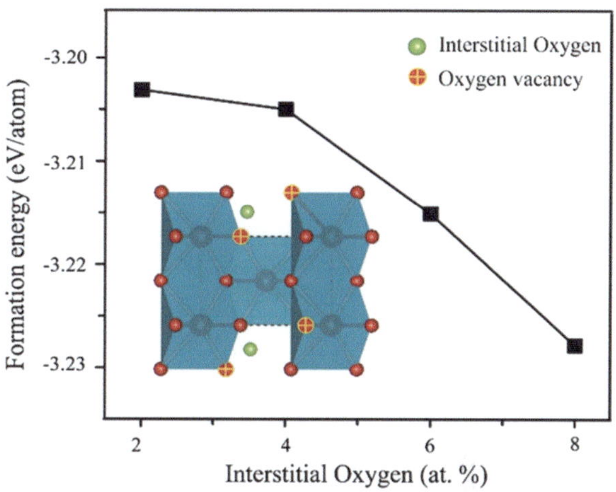

Fig. 6.13 T-phase formation energy for an O vacancy of 10% varies with the introduction of interstitial oxygen

coating in a simulated pressurized water reactor environment and assessed the corrosion resistance of the coating before and after irradiation. Figure 6.14 shows the XRD spectrum changes of ZrO_2 coatings with 0 DPA and 1.2 DPA after oxidation. The (101) peak of T–ZrO_2 corresponding to the original coating (0DPA) completely disappeared after 40 h of oxidation. As a comparison, the irradiated (1.2DPA) coating showed good tetragonal stability. Figure 6.15 shows the additional increase in oxide thickness after different oxidation times. The T–ZrO_2 phase becomes more stable after irradiation, and compared with the non-irradiated coating, the irradiated modified coating exhibits higher oxidation resistance. Observation of the surface and cross-sectional morphology of the coating using SEM (Fig. 6.16) showed that the surface of the original ZrO_2 became very rough after oxidation, while the surface of DPA1.2 remained smooth and intact.

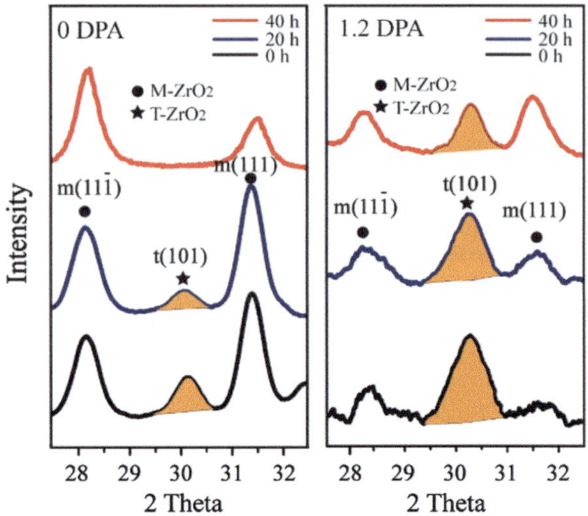

Fig. 6.14 Comparison of XRD spectral changes of ZrO_2 coatings with and without irradiation after different times of oxidation

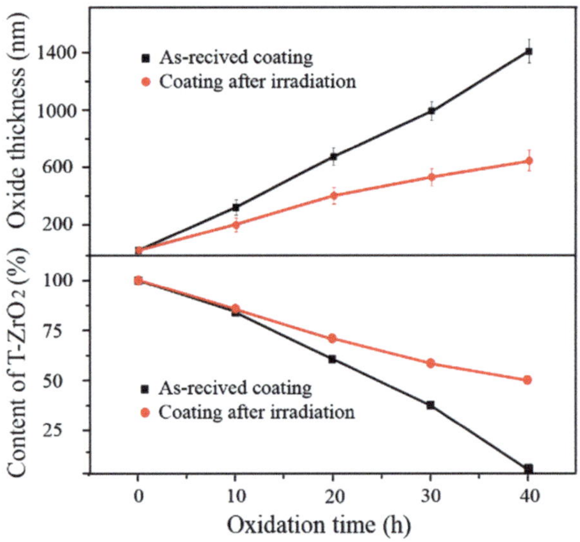

Fig. 6.15 Oxidation behavior of oxide coatings with and without radiation. Retention of T–ZrO_2 shows the same kind of variation

The formation of the T-phase will be further consolidated considering that the irradiation effect is continuous in real reactors. Figure 6.17 shows the growth conditions of oxide coatings with different irradiation doses and different oxidation times. We can see that the oxidation kinetics is suppressed with increasing irradiation dose. The stability of T–ZrO_2 enhances the oxidation resistance of the coating. During oxidation, the formed zirconia initially presents a mixture of tetragonal and monoclinic ZrO_2 crystal structures. The formation of the tetragonal phase is mainly due to

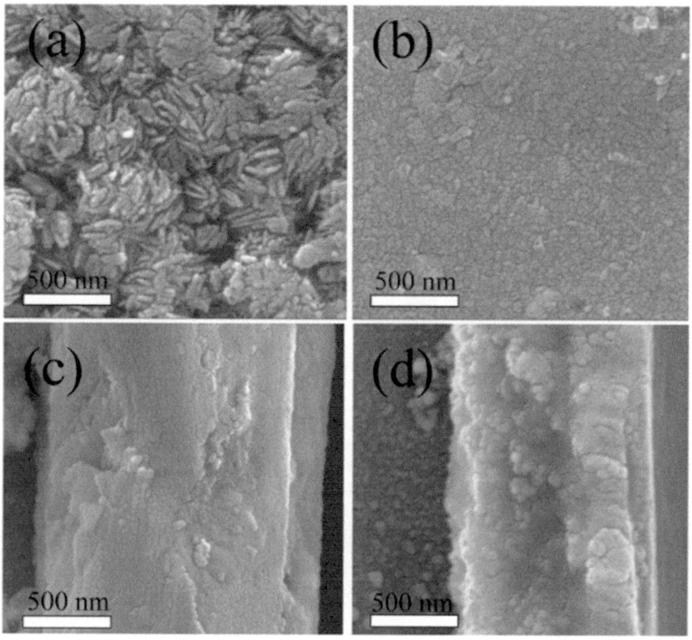

Fig. 6.16 Surface morphology (**a**, **b**) and cross-section (**c**, **d**) morphology of ZrO_2 thin films after 40 h oxidation: under 1.2 DPA irradiation (**b**, **d**) and unirradiated (**a**, **c**)

the smaller grain size and internal stresses. However, as the oxide grows, the stresses are gradually released. At the same time, the transformation of the tetragonal phase to the monoclinic phase can be facilitated by an increasing grain size above a critical value. As a result, a surface collapse caused by the volume change of the phase transition can be observed (Fig. 6.16). Naturally occurring ZrO_2 may be unprotected due to volume changes (phase transitions) and cracking, which provides an easy way to diffuse the material.

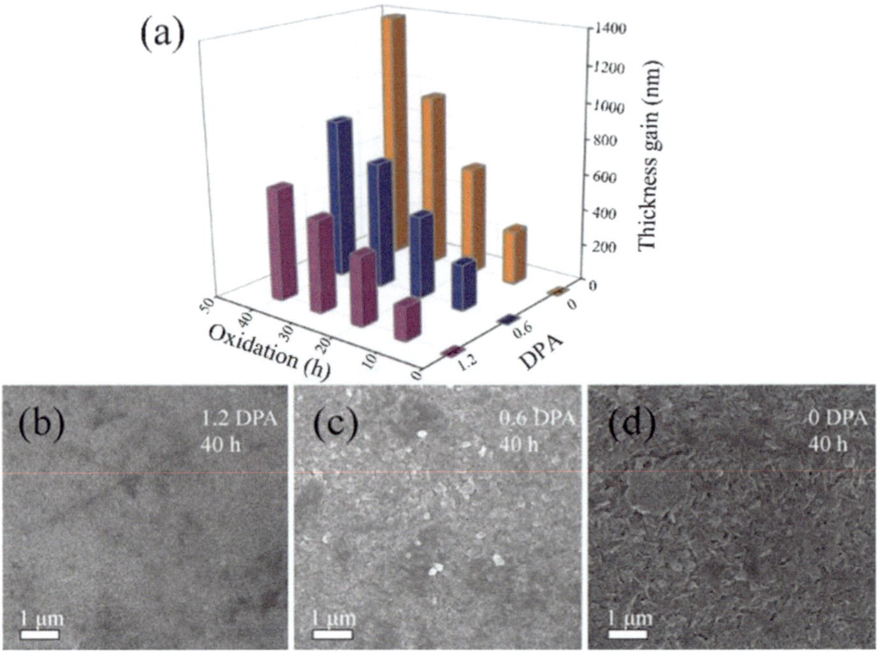

Fig. 6.17 Increase in thickness of oxide film and surface morphology of oxide film at different irradiation doses and different oxidation times

6.3 Conclusion

In summary, the special pathway of ion irradiation modification not only promotes the transformation from monoclinic phase to tetragonal phase, but also stabilizes it at room temperature. The synergistic effect of T/M and irradiation-induced oxygen vacancy gap on the interface gradient mismatch promotes the stability of T–ZrO_2, thereby improving the oxidation resistance of the coating. Meanwhile, radiation-induced Zr–O bond contraction will further inhibit oxygen diffusion. Therefore, the project proposes a method to improve corrosion resistance through a self-stabilizing mechanism. Under the actual conditions of the reactor core, oxidation and ion bombardment occur simultaneously. We cannot distinguish the effects of oxidation and irradiation on coating materials separately. However, this work provides ideas for the application of coatings inside reactors and has guiding significance for extending the service life of nuclear cladding.

ZrO_2 nanocrystalline coatings with controllable oxygen defects were prepared using the HiTUS system with zirconium alloy as the substrate under different process parameters. It was demonstrated that PLS power, target bias pressure, and oxygen partial pressure affect the microstructure of the coatings. The irradiation behavior of nanocrystalline ZrO_2 coatings containing M and T-phases was investigated using Ar^{8+} bombardment. Irradiation was found to further introduce a large number of oxygen defects and lattice distortions, which led to the enhancement of Zr–O bonds and increased the elastic modulus of the material. The T-phase became the dominant phase and was further stabilized after irradiation by vacancy-gap pairs. The anomalous martensitic transformations under Ar^{8+} irradiation were induced by gradient mismatches in the vicinity of lattice distortion regions. The irradiated samples were found to have a very high tetragonal phase retention (about 57%) after oxidation, and the coating remained overall dense and homogeneous without large cracks. In contrast, the samples in the as-deposited state underwent a complete transition to the monoclinic phase and the oxide film thickness increased considerably.

References

1. Hu J, Yang L, Cao G, Yun Y, Yuan G, Yue Q, Shao G (2016) On the oxidation behavior of $(Zr, Nb)_2Fe$ under simulated nuclear reactor conditions. Corros Sci 112:718–723
2. Duan Z, Yang H, Satoh Y, Murakami K, Kano S, Zhao Z, Abe H (2017) Current status of materials development of nuclear fuel cladding tubes for light water reactors. Nucl Eng Des 316:131–150
3. Yuan G, Cao G, Yue Q, Yang L, Yun Y, Shao G, Hu J (2016) Formation and finestructures of nano-precipitates in ZIRLO. J Alloys Compd 687:451–457
4. Cao G, Yun Y, Yang L, uan G, Yue Q, Shao G, Hu J (2017) The formation and stacking faults of Fe and Cr containing Laves phase in Zircaloy-4 alloy. Mater Lett 191:203–205 (2017)
5. Motta T, Arthur C, Adrien JC (2015) Robert, corrosion of zirconium alloys used for nuclear fuel cladding. Annu Rev Mater Res 45:311–343
6. Hulme H, Baxter F, Babu RP, Denecke MA, Gass M, Steuwer A, Preuss M (2016) An X-ray absorption near-edge structure (XANES) study of the Sn L3 edge in zirconium alloy oxide films formed during autoclave corrosion. Corros Sci 105:202–208
7. Bell BDC, Murphy ST, Burr PA, Comstock RJ, Partezana JM, Grimes RW, Wenman MR (2016) The influence of alloying elements on the corrosion of Zr alloys. Corros Sci 105:36–43
8. Ribis J, Wu A, Brachet JC, Barcelo F, Arnal B (2018) Atomic-scale interface structure of a Cr-coated Zircaloy-4 material. J Mater Sci 53:9879–9895
9. Wu J, Ribis JC, Brachet E, Clouet F, Leprêtre E, Bordas B (2018) Arnal, HRTEM and chemical study of an ion-irradiated chromium/Zircaloy-4 interface. J Nucl Mater 504:289–299
10. Michau F, Maury F, Schuster F, Lomello JC, Brachet E, Rouesne M (2018) Pons, High-temperature oxidation resistance of chromium-based coatings deposited by dli-MOCVD for enhanced protection of the inner surface of long tubes. Surf Coat Technol 349:1048–1057
11. Lin K, Hwang M, Wu C (1996) The deposition and wear properties of cathodic arc plasma deposition TiAlN deposits. Mater Chem Phys 46:77–83
12. Y. Chen, X. Nie, D.O. Northwood, Investigation of Plasma Electrolytic Oxidation (PEO) coatings on a Zr-2.5 Nb alloy using high temperature/pressure autoclave and tribological tests, Surf. Coat. Technol. 205 (6) (2010) 1774–1782.
13. Hobbs LW, Rosen VB, Mangin SP, Treska M, Hunter G (2005) Oxidation microstructures and interfaces in the oxidized zirconium knee. Int J Appl Ceram Tec 2(3):221–246
14. Yang J, Wang X, Wen Q, Wang X, Wang R, Zhang Y, Xue W (2015) The effect of microarc oxidation and excimer laser processing on the microstructure and corrosion resistance of Zr-1Nb alloy. J Nucl Mater 467:186–193
15. Wang L, Hu X, Nie X (2013) Deposition and properties of zirconia coatings on a zirconium alloy produced by pulsed DC plasma electrolytic oxidation. Surf Coat Technol 221:150–156
16. Simeone D, Gosset D, Bechade JL, Chevarier A (2002) Analysis of the monoclinic–tetragonal phase transition of zirconia under irradiation. J Nucl Mater 300(1):27–38

17. Verlet R, Tupin M, Baldacchino G, Wolski K, Miro S, Gosset D, Colas K, Jublot M, Jomard F (2015) Influence of light ion irradiation of the oxide layer on the oxidation rate of Zircaloy-4. Corros Sci 98:327–338
18. Panigrahy M, Aslam D (2011) Bahadur, Controlled optical and magnetic properties of ZnO nanorods by Ar ion irradiation. Appl Phys Lett 98(18):183109
19. Wakeham S, Thwaites M, Tsakonas C, Cranton W, Ranson R, Boutaud G, Koutsogeorgis D (2010) A new reactive sputtering technique for the low temperature deposition of transparent light emitting ZnS: Mn thin films. Physica Status Solidi A Appl Res 207:1614–1618

Chapter 7
Molecular Designing of ZSO Coatings to Achieving "Diffusion Barrier"-"Hydrochemical Stability" Trade-Off

Abstract Silicon-contained coatings are rarely used on Zr alloy cladding tubes, mainly because the continuous SiO_4 tetrahedral structure is easy to be destroyed in subcritical environment. Based on the continuous mixing structure of SiO_x tetrahedron and ZrOx polyhedron, amorphous Zr–Si–O (ZSO) coating was proposed in this work. The key to the long-term protection of ZSO coating lies in the bonding status and structural stability under working conditions. According to first-principle calculation and atomic bonding characterization, the stable mechanism of amorphous ZSO coating was established, from the point of elemental diffusion and the stability of two interfaces (Zr/ZSO, ZSO/H_2O). Based on the regulation of atomic bonding and packing condition, a trade-off of "diffusion barrier" and "hydrochemical stability" was designed in the optimized $ZrSi_3O$ coating experimentally. All the elements were in a strong hybrid bonding state, which ensures the stability of atomic structure and long-term protection performance. The results of the 400 °C air environment oxidation experiment show that the Zr–O–Si coating can provide good protection for Zr alloys. When subjected to air burning oxidation at 600 and 800 °C, the coating still provides protection to the Zr alloy, but microcracks begin to appear on the surface of the coating.

Keywords Amorphous Zr–Si–O · First-principle calculation · Bonding status · Diffusion barrier · Hydrochemical stability

7.1 Introduction

Zirconium alloy has a small thermal neutron absorption cross-section, high thermal conductivity, good mechanical properties, and good compatibility with uranium dioxide, making it the perfect cladding material for nuclear power plants since the 1960s [1]. Under normal working conditions, the main degradation problems of Zr alloys in contact with coolant are related to general oxidation, nodular corrosion, galvanic corrosion, shadow corrosion, localized corrosion caused by scale accumulation, hydrogenation, and debris micro motion [2, 3]. The zirconium alloy with solid

J. Hu and G. Cao, *Zirconium Alloy Coatings*,
https://doi.org/10.1007/978-981-96-5028-6_7

solution of oxygen or hydrogen is weakened in mechanical properties and is prone to corrosion induced cracking [4]. The in situ formed zirconia layer can be regarded as a passivation film, which indeed improves the corrosion resistance. However, the phase transition from tetragonal to monoclinic zirconia generated obvious volume expansion, and the resulting void and cracks are short diffusion paths for corrosive media [5].

The introduction of coatings is the most effective approach for antioxidation performance in short term [6]. At present, the coating types are mainly metallic Cr or alloy systems [7]. Due to the weak bonding nature of metal bonds and the large diffusion coefficient of metal atoms in zirconium and zirconia, metallic coatings are faced with the problems of interfacial reaction and element migration. Although the formation of interfacial phase is beneficial to metallurgical bonding, the continuous diffusion gradually reduces the structural integrity and continuity of the protective layer. Ceramic coating has high elastic modulus. The main bond between the atoms is covalent, and the bond energy is strong. During our preliminary research on Zr–Si coating system, the oxidation resistance is closely related to the Zr–Si–O (abbr.: ZSO) amorphous oxide layer formed on the surface and interface with superior oxygen-barrier performance [8–10] Si has a very low thermal neutron absorption cross-section and strong amorphous forming ability. The research on amorphous ZSO coating is a cutting-edge work. The arrangement of atoms from long-range order to short-range disorder can improve the shortcomings of the brittleness of oxide materials and improve the viscoelasticity of the coating/substrate interface [11, 12]. In addition, the number of charged oxygen vacancies in zirconium-oxygen polyhedron is reduced due to the trapping of electrons in Si atoms, which leads to a decrease in the mobility of oxygen ions in the oxide layer [13, 14]. However, Si-based coating could not maintain sufficient structure integrity since SiO_x as the oxidation product dissolves easily in hot water [15]. The Zr silicide coating fabricated by Yeom et al. exhibited significant weight loss during the corrosion test in pure steam at 400 °C [16]. Silicon depletion and the formation of porous ZrO_2 layer on $ZrSi_2$ surface were recognized as the main reason for the surface recession. This fatal weakness limits the application of silicon-based coatings in light-water reactor environments.

In order to apply silicon containing coatings with potential advantages over other materials, new strategies should be adopted to address the challenges of stability under hydrothermal conditions. In this work, based on first-principle calculation and atomic bonding characterization, the stability of amorphous ZSO coating was discussed, from the point of elemental diffusion and the stability of two interfaces (Zr/ZSO, ZSO/H_2O). Amorphous ZSO coatings were sputtered on Zr substrates with metallurgical interface bonding. The protective effect is extremely sensitive to the local atomic packing. A trade-off of "diffusion barrier" and "hydrochemical stability" was obtained.

7.2 Experiment

ZSO coatings were deposited by magnetron sputtering on different substrates: single crystalline Si(110) with deposited Zr-4 layer and Zr-4 alloy. For the substrate preparation, the Si wafers were ultrasonically cleaned in ethanol and acetone for 15 min to obtain a smooth and clean surface without contaminants. Then, a Zr-4 layer was deposited by direct current (DC) power. The Zr alloy substrate was ground by 600# sandpaper and ultrasonically cleaned in acetone and ethanol for 10 min.

A three-inch Zr-4 target (purity 99.95%) and silicon target (purity 99.99%) were used during the deposition process. The ZSO films were deposited by co-sputtering with oxygen as reactive gas. Si and Zr-4 targets were driven by radiofrequency (RF) generators and DC power, respectively. A gas mixture of Ar and O_2 was introduced. The flow rate of argon and oxygen was 20 sccm and 4.5 sccm, respectively. The ZSO films were named by $ZrSi_xO$, where X represents the atomic ratio of Si to Zr.

Oxidation tests were conducted in an industrial furnace at 400 °C in ambient air with different oxidation times. A temperature of 400 °C is sufficient to study the structural stability and oxidation behavior of multilayer structures under normal reactor conditions. Then, the optimized Zr/Si multilayer film was tested for different durations in a static autoclave (Anhui Komi Machinery Technology Co., Ltd.) in high-temperature and high-pressure water (HHW) at 320 °C and 15.5 MPa. The surface and cross-sectional morphology as well as chemical composition were characterized using a field-emission scanning electron microscope (FE-SEM, JEOL, Japan) equipped with energy-dispersive spectroscopy (EDS). Transmission electron microscopy (TEM, FEI Tecnai G^2 F20) was used to observe the microstructure of the coating and the thickness of the zirconia layer. Cross-sectional samples were prepared by ion beam thinning using 4 keV Ar ions on Gantan 691. In order to reduce the damage of TEM thin films caused by ion milling, ion milling is performed at an angle of 3–5°. EELS was collected using a transmission electron microscope equipped with a field-emission gun (FEG) and operated at 200 keV using a Gatan 965 PEELS spectrometer and ELP data acquisition system.

AIMD calculation was used to evaluate oxygen diffusion behavior in Zr–Si–O amorphous phase using VASP [17]. Amorphous structure was obtained by heating corresponding crystals of different compositions to 5000 K and then rapidly quenching them to induce amorphous structure. During the calculation process, after pre-balancing for 10 ps, perform each AIMD task on a traditional unit containing 96 atoms within 90 ps. A time step of 2 fs was used in the standardized ensemble [18].

7.3 Results and Discussions

The ZSO coatings with varied atomic ratio of Si/Zr were deposited in this work. The microstructure was analyzed by TEM. For $ZrSi_{0.26}O$, the bright-field image shows obvious diffraction contrast in Fig. 7.1. The enlarged image shows the formation

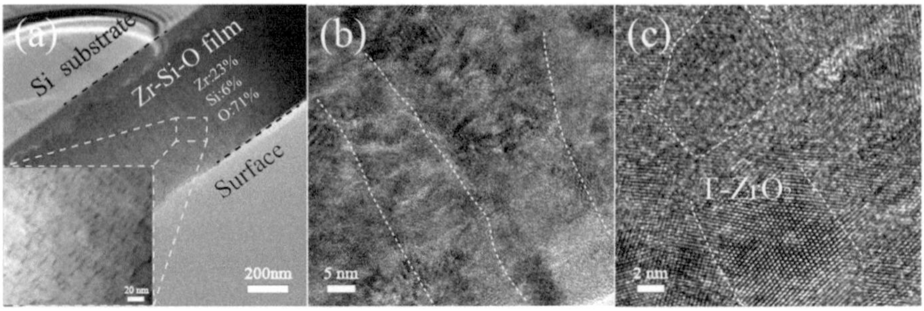

Fig. 7.1 Microstructure of as-deposited $ZS_{0.26}O$ film

of a discrete columnar crystal structure. The HRTEM confirmed the formation of T-ZrO_2 in Fig. 7.1c. Further increase of Si in ZSO resulted in the formation of amorphous structure. The amorphization effect of Si has been reported in ZrO_2-SiO_2 system [19]. The absence of grain boundaries and other lattice defects in amorphous coatings could reduce the number of short-circuit diffusion channels to some extent, thus promoting chemical and structural homogeneity during corrosion. In order to maintain a homogeneous amorphous structure, the Si:Zr ratio in ZSO system was maintained between 0.4 and 4 in this work. To identify the atomic modification effect of Zr on Si–O bonds, the SiO_2 coating was deposited for comparison.

Figure 7.2a–d shows the cross-sectional TEM images of as-deposited $ZrSi_1O$, $ZrSi_3O$, $ZrSi_4O$, and SiO_2. The selected area electron diffraction (SAED) patterns show obvious broad halo ring, indicating that the ZSO coatings and SiO_2 are in amorphous state. The diameter of the amorphous ring can represent the average nearest neighbor interatomic distance [20]. These values for $ZrSi_1O$, $ZrSi_3O$, $ZrSi_4O$, and SiO_2 were 6.645 1/nm, 6.791 1/nm, 6.818 1/nm, and 6.845 1/nm, respectively. The average interatomic distance of amorphous ZSO was reduced due to the increase of Si content.

To verify the existence of specific bonding states, the atomic structure of as-deposited ZSOs and SiO_2 was further studied by EELS. The examination of O K in Fig. 7.2e suggests that the introduction of Zr changes the atomic coordination of oxygen. The onset of O K-edge for SiO_2 is 531.5 eV. Addition of Zr shifts the O K-edge onset to 529.9 eV in $ZrSi_1O$. The presence of Zr further affects the Si $L_{2,3}$-edge onset, which will be discussed latter. The O K-edge of ZSO demonstrated a complex structure with the increase of Zr content. In the ZSO sample, the structure can be divided into two different regions. The first one is related to the transition toward empty states dominated by Zr 4d, which hybridize with oxygen 2p orbitals. The function of adjacent Zr atoms is to split the p-type DOS of oxygen into t_2 and e_g energy levels through the crystal field splitting of Zr's 4d orbitals [21]. The second region should be related to the transition to a state mainly formed by hybridization of silicon 3s/3p/3d orbitals with oxygen 2p orbitals, although the influence of zirconium 5s and 5p orbitals has not been ruled out [22].

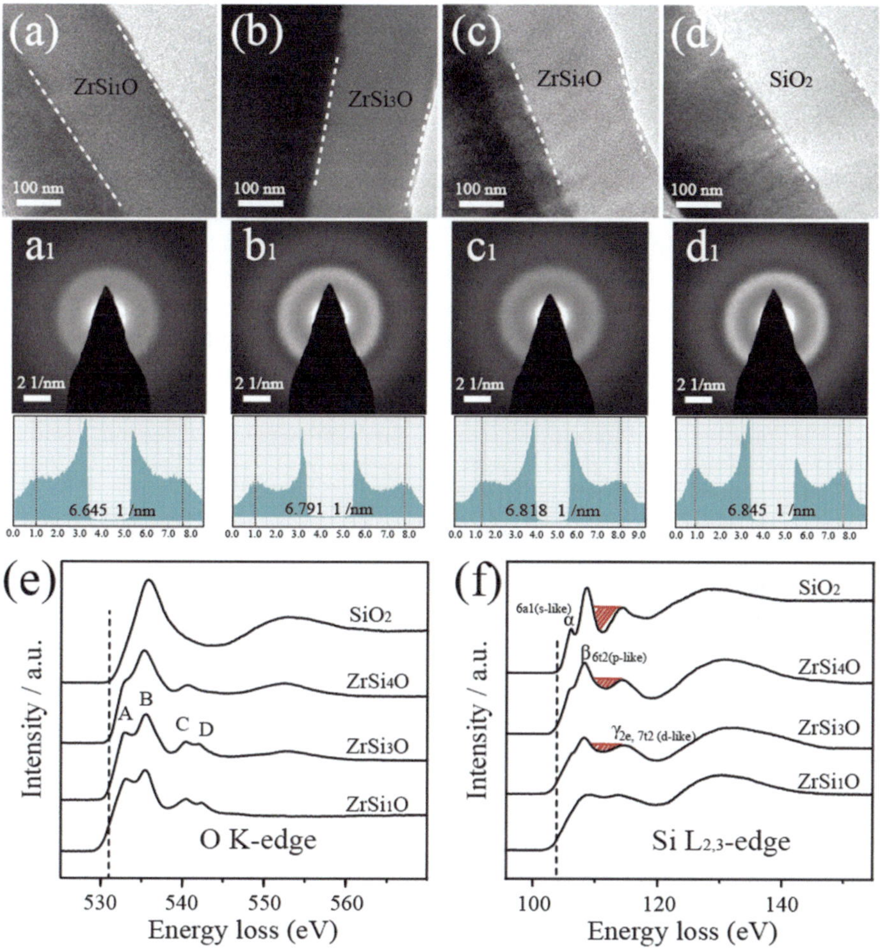

Fig. 7.2 **a–d** Cross-section TEM image and corresponding selected area electron diffraction pattern of $ZrSi_1O$, $ZrSi_3O$, $ZrSi_4O$, and SiO_2. The bottom shows the measurement of the amorphous diffraction rings. **e**, **f** EELS results of O K-edge and Si L2,3-edge for $ZrSi_1O$, $ZrSi_3O$, $ZrSi_4O$

The Si $L_{2,3}$ core-loss edge was used to probe the chemistry difference around Si atoms in SiO_2 and ZSO systems (Fig. 7.2f). The Si $L_{2,3}$-edge exhibits three main features that can be assigned to transitions from the 2p core level to the $6a_1$(s-like), $6t_2$(p-like), 2e(d-like), and $7t_2$(d-like) of an SiO_4 cluster [23]. The spectrum of $ZrSi_4O$ is almost identical to that of SiO_2. With the increase of Zr content, the edge onsets of peaks are shifted slightly toward lower energies, which correspond to a decrease in effective positive charge of Si atoms [24]. The Si $L_{2,3}$-edge is affected by coordination of the next-nearest-neighbor cations in Si–O–Zr. Since Zr^{4+} is a small and highly polarizing ion, the introduction of Zr^{4+} would result in modification of the electron density in SiO_x tetrahedron. The energy of the $6a_1$ state in molecular orbitals is highly sensitive to electronic changes, while the remaining quasi-bound states are not [25]. An increase in electron concentration around silicon would cause a corresponding shift to higher energy of the bound state, $6a_1$, decreasing the separation between the $6a_1$ and $6t_2$ molecular orbitals. Indeed, as shown in the Si $L_{2,3}$-edge of ZS_3O, peaks of α and β has almost merged.

The shape of the spectrum also changed. If the structure of ZSO was dominated by micro domains of SiO_2 and ZrO_2, then the Si $L_{2,3}$-edge would closely resemble the edge of amorphous SiO_2. The differences between the edges of ZSO and amorphous SiO_2 are consistent with a structure composed of a random network of SiO_x and ZrO_x polyhedral for ZSO system. The shape variation of Si $L_{2,3}$-edge can be qualitatively explained for in terms of symmetry change of molecular orbital and distortion of SiO_x polyhedral in ZSO. The polyhedral distortion would broaden peaks due to the large range of bond distances and angles and hence to a widening of the energy bands [24]. The contribution of transitions to the $6t_2$ final state would also be split, resulting in the filling-in between β and γ in ZSO, different from the spectra of non-distorted tetrahedral in SiO_2, as marked in the Si $L_{2,3}$-edge spectra.

The redox reaction between oxides and Zr has been reported in previous Cr_2O_3/Zr system. To obtain a stable ZSO structure, it is necessary to have an inert ZSO/Zr interface. In addition, in high-temperature and high-pressure water (HHW) environment, SiO_2 would react with H_2O severely, resulting in the formation of $SiO(OH)^{-3}$, which were easily dissolved [26]. In order to evaluate the stability of ZSO/Zr and ZSO/H_2O interfaces, amorphous $ZrSi_1O$, $ZrSi_3O$ and $ZrSi_4O$ models were constructed adjacent to three layers of Zr atoms by AIMD and relaxed. In Fig. 7.3a, the total radial distribution function (RDF) proved a more compact atomic arrangement in $ZrSi_4O$. As a result, the free volume in ZSO remarkably decreased which helps to eliminate the oxygen diffusion pathways. The interface reaction and the diffusion of oxygen for $ZrSi_1O$/Zr and $ZrSi_4O$/Zr couples are illustrated in Fig. 7.3b. One can see that large amount of oxygen in $ZrSi_1O$ was diffused into the modeled Zr alloy. This is mainly because zirconium is more oxygen-friendly. As comparison, the oxygen in $ZrSi_4O$ was quite stable. The loss of oxygen in amorphous ZSO means the collapse of the Zr–O–Si atomic packing structure. The increased silicon content helps to stabilize the structure of ZSO. For systems with low silicon content, zirconium-oxygen bonding dominates, and there may be some free or weakly bonded forms of Si–O inside.

In Fig. 7.3c, some H_2O molecules were put on the surface of $ZrSi_1O$, $ZrSi_3O$ and $ZrSi_4O$ and relaxed by VASP. The adsorption and decomposition of water molecules on the surface of ZSO at high temperature were investigated. The bonding between hydroxyl and silicon was observed on the surface of $ZrSi_4O$, but no obvious bonding relationship was found in $ZrSi_1O$. This should be related to the state of charges around silicon atoms. When the silicon-oxygen bond is weak (that is, when the electron density around the silicon atom is high), the tendency to dissolve in the reaction with subcritical water is small. Both interfaces of $ZrSi_1O$/H_2O and $ZrSi_1O$/Zr reached a stable level. One can see that Si–O bonds tend to be adjacent in $ZrSi_4O$ in Fig. 7.3d. For the –Si–O–Si–O– bonding, oxygen 2p orbitals can interact significantly with silicon 3s/3p and 3d orbitals. Silicon has a fairly positive charge (with a lower

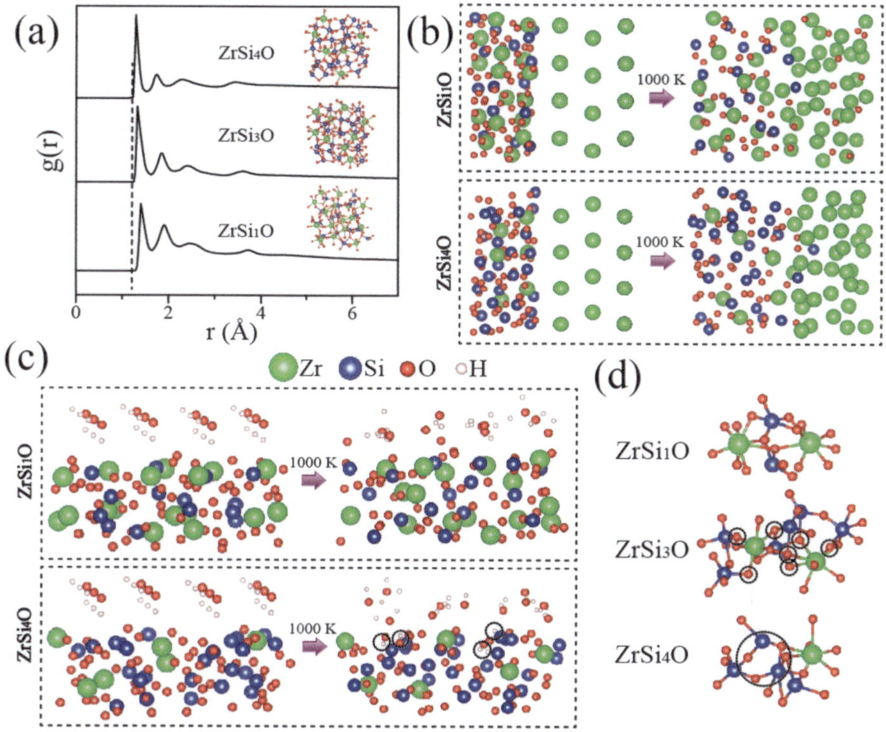

Fig. 7.3 **a** Total RDF of $ZrSi_1O$, $ZrSi_3O$, and ZrSi4O. The atomic modals were inserted. **b** The AIMD simulation of the interface behavior for $ZrSi_1O$/Zr and $ZrSi_4O$/Zr at 1500 K. The oxygen diffusion from ZSO to Zr was largely restricted in $ZrSi_4O$ system. **c** The AIMD simulation of the interface behavior for $ZrSi_1O/H_2O$ and $ZrSi_4O/H_2O$ at 1000 K. The water molecules are more likely to adsorb and decompose on the surface of $ZrSi_1O$. **d** The atomic structures of $ZrSi_1O$, $ZrSi_3O$, and $ZrSi_4O$, which were constructed and relaxed by VASP

electron density surrounded). The OH^- dissociated from water is more likely to adsorb at the Si site with a large positive charge. Thus, a large number of –Si–O–Si–O– bonds in $ZrSi_4O$ are prone to dissolution. The presence and increase of Zr can break this bonding form. The special bonding mode of Si–O–Zr also increases the charge density around the silicon atoms to a certain extent, as shown by the EELS results, reducing the adsorption tendency of OH^-. Thus, Zr plays a special role in holding back the reaction with water.

The protective performance of ZSO coatings was further analyzed by experiment. The oxygen-barrier properties and the interface stability of the coating and matrix were investigated at 400 °C. Based on the above theoretical calculation, four coatings of $ZrSi_{0.6}O$, $ZrSi_1O$, $ZrSi_{1.3}O$, and $ZrSi_{1.8}O$ were introduced. As shown in Fig. 7.4a, b, Zr alloy beneath the coatings oxidized after 100 h oxidation. For $ZrSi_{1.3}O$, the oxygen-barrier property increased. The form ZrO_2 layer was quite thin after 200 h oxidation. For $ZrSi_{1.8}O$, no ZrO_2 layer was found after 200 h oxidation. It also need to

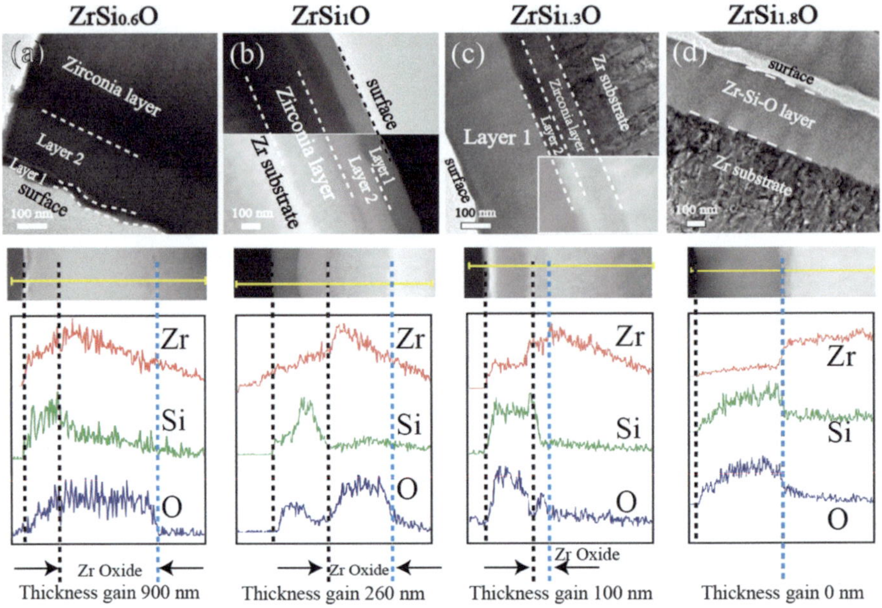

Fig. 7.4 **a**, **b** Cross-sectional macrostructure of $ZrSi_{0.6}O$, $ZrSi_1O$ after 100 h oxidation at 400 °C. **c**, **d** The cross-sectional macrostructure of $ZrSi_{1.3}O$, and $ZrSi_{1.8}O$ after 200 h oxidation at 400 °C. The bottom shows the EDS line-scan results. The thickness gain of the oxide layer for the underneath Zr alloy was also shown

mention that $ZrSi_{0.6}O$, $ZrSi_1O$ and $ZrSi_{1.3}O$ demonstrated stratification phenomenon. A transition layer (labeled as layer 2) appeared between top ZSO layer (labeled as layer 1) and newly formed ZrO_2 layer.

A comprehensive evaluation of ZSO coatings was carried out to understand the diversity microstructural changes and the oxygen-blocking efficiency at 400 °C. Figure 7.5 demonstrates the fine structures of oxidation changes for $ZrSi_{0.6}O$, $ZrSi_1O$, and $ZrSi_{1.3}O$ after oxidation for 100 h. The ZSO coatings have transformed into two-layered structures (Fig. 7.5a, e, h). For $ZrSi_{0.6}O$, the thickness of layer 1 was quite small while that of Layer 2 was large. The oxygen content in Layer 2 decreased to 50 at. %. Layer 2 was in amorphous state. The Si content decreased in Layer 1, which triggered the crystallization. $ZrSi_{0.6}O$ possessed poor oxygen diffusion resistance, since the underneath ZrO_2 layer kept increase to about 900 nm. The microstructure evolution of $ZrSi_1O$ after oxidation at 400 °C was shown in Fig. 7.5e–g. Similar to $ZrSi_{0.6}O$, the ZSO layer also has transformed into a two-layered structure. However, the inward diffusion of Si in $ZrSi_1O$ was restrained. The Si:Zr ratio in Layer 1 and Layer 2 is 0.96:1 and 0.97:1, respectively, which is almost unchanged from the original sample. The HRTEM image in Fig. 7.5g shows that Layer 1 and Layer 2 remained in amorphous state. The O content in Layer 1 and Layer 2 of $ZrSi_1O$ is 68 at. % and 22 at. %, respectively. This is the reason for mass thickness contrast in Fig. 7.5g. The thickness of the underneath ZrO_2 layer in $ZrSi_1O$ is 260 nm, which is smaller than that of $ZrSi_{0.6}O$. For $ZrSi_{1.3}O$, as shown in Fig. 7.5h–j, the thickness of Layer 2 is only 56 nm and the formed ZrO_2 layer is also largely reduced (95 nm), indicating that the interface reaction and element diffusion have been further weakened. The formation of the transition layer with less oxygen effectively inhibits the further occurrence of interface reaction and element diffusion.

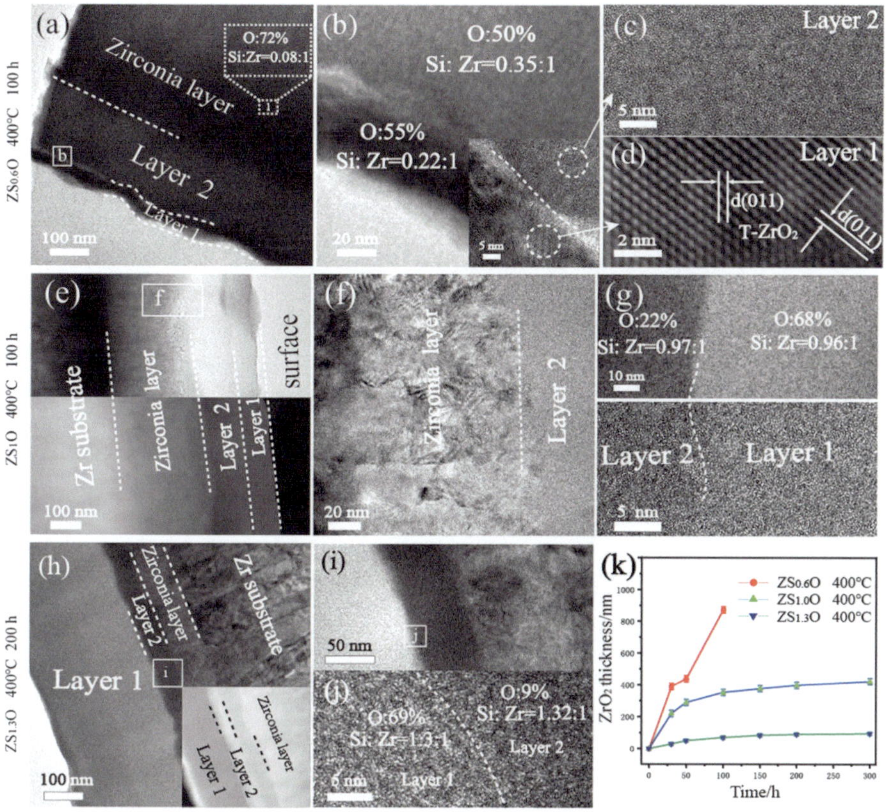

Fig. 7.5 TEM images of the cross-sectional information for $ZrSi_{0.6}O$ (**a–d**), $ZrSi_1O$ (**e–g**) and $ZrSi_{1.3}O$ (**h–j**)-coated Zr alloys after oxidation at 400 °C for 100 h. **a** The bright-field image for ZrSi0.6O after 100 h oxidation. **b** The enlarged image from the marked rectangular box in Fig. 7.5a. The insert image demonstrated the interface between Layer 1 and Layer 2. **c**, **d** HRTEM images for Layers 1 and 2 after 100 h oxidation. **e** The bright-field and HAADF image for ZrSi1O after 100 h oxidation. The thickness of each layer was almost unchanged. **f** The enlarged image from the marked rectangular box in Fig. 7.5e. **g** HRTEM image of Layer 1 and Layer 2 with EDS analysis **h** The bright-field image of ZrSi1.3O.with each layer marked after oxidation at 400 °C after 200 h. The insert is the corresponding HAADF image. **i** The enlarged image from the marked rectangular box in Fig. 7.5h. **j** HRTEM image from the marked rectangular box in Fig. 7.5i. **k** Oxidation kinetics curves of different ZrSiO coatings

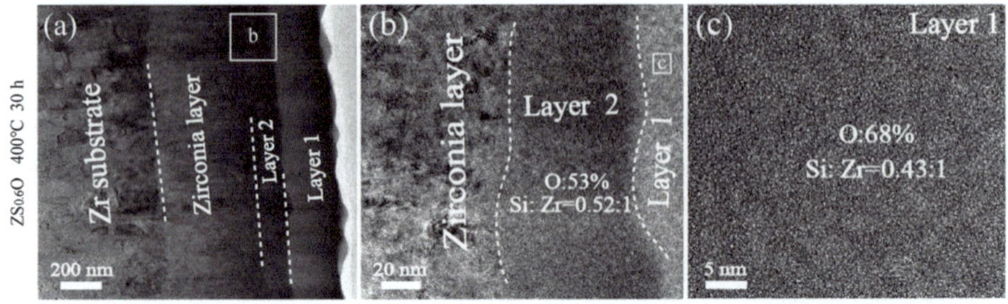

Fig. 7.6 TEM images of the cross-sectional information for $ZrSi_{0.6}O$-coated Zr alloy after oxidation at 400 °C for 30 h (**a–c**) and 100 h (**d–f**). **a** The bright-field image after 30 h oxidation. The formation of zirconia layer was attributed to the poor barrier property of $ZrSi_{0.6}O$. **b** Enlarged image from the marked rectangular box in Fig. 7.5. The $ZrSi_{0.6}O$ transformed into a two-layered structure. **c** HRTEM image of Layer 1 with EDS analysis

To further understand the phase transition in layered $ZrSi_{0.6}O$, the microstructure evolution in early oxidation stage was shown in Fig. 7.6. After oxidation for 30 h, the two-layered structure has already formed with thicker Layer 1 and thinner Layer 2. Both Layers 1 and 2 remained in amorphous state. Layer 2 was in oxygen-depleted state with O content of 53 at. %, while Layer 1 maintained the complete oxidation state. The formation of Layer 2 is attributed to the interface reaction between $ZS_{0.6}O$/Zr and the inward diffusion of O in Layer 1. Thermodynamically, Zr can take the oxygen out of SiO_2 with the formation of ZrO_2 at 400 °C. The decrease of oxygen in Layer 2 is due to the continuous diffusion of O into the Zr through the interface reaction. The Si content in Layer 1 was smaller compared to that in Layer 2. This confirms that the silicon in the surface ZSO has a tendency to diffuse further inward during oxidation in $ZrSi_{0.6}O$.

Figure 7.7 shows the thickness and Si/Zr ratio changes of Layers 1 and 2 in $ZrSi_{0.6}O$ (0–100 h). One can see that Layer 1 gradually transformed into Layer 2,

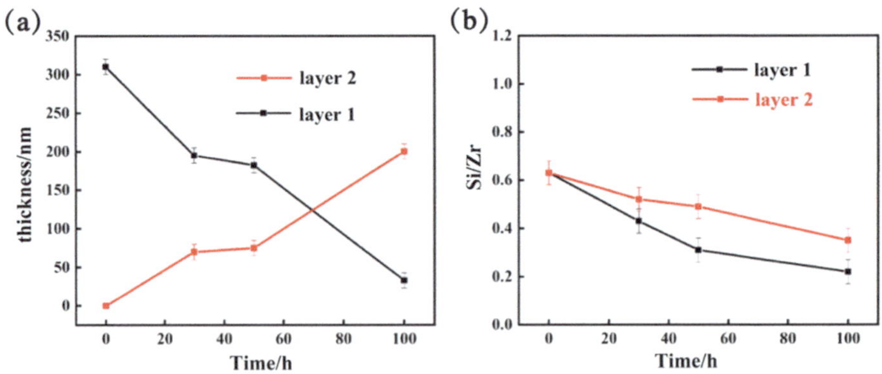

Fig. 7.7 Statistical analysis for $ZrSi_{0.6}O$ after oxidation at 400 °C. **a** The thickness change of Layer 1 and Layer 2 in $ZrSi_{0.6}O$ with oxidation time. **b** Si/Zr ratio changes with oxidation time

indicating an oxygen transfer process. Layer 2 derived from the de-oxidation of Layer 1, and the oxygen in Layer 1 diffused into Zr alloys. The Si/Zr ratio of Layers 1 and 2 were both reduced, and the internal diffusion of silicon in Layer 1 is more severe. The thickness variation of Layers 1 and 2 in $ZrSi_1O$ is also obtained in Fig. 7.8, as well as the Si/Zr ratio and O content changes. Unlike $ZrSi_{0.6}O$, the thickness change of Layers 1 and 2 is negligible from 100 to 200 h. In addition, the Si:Zr ratio in Layers 1 and 2 has not changed with prolonged oxidation, and both maintained in amorphous state. The oxygen content in Layer 2 further decreased to 10 at. %. It is plausible to say that the O source for the formation of ZrO_2 mainly comes from interface reaction of ZSO/Zr in initial stage (0–30 h). As the oxidation time is prolonged, the O from Layer 2 gradually stops supply.

In ZSO coatings, when the element zirconium is dominant, the Zr-O bonds and the ZrO_7 polyhedron in the amorphous structure are dominant. At this time, there are some silicon and oxygen ions that are not tightly bonded, and the dissociation energy of silicon and oxygen is small. The oxygen-induced internal diffusion of silicon was occurred. At the same time, oxygen that is not bonded tightly diffused into the zirconium matrix. Figure 7.9 shows the structure evolution of $ZS_{0.4}O$ after oxidation at 400 °C for 30 h and 100 h. The Si inner diffusion was quite obvious, finically leading to a loose structure. As comparison, when silicon is dominant, the Si-O bond from tetrahedron-dominated networks formed in ZSO has large dissociation

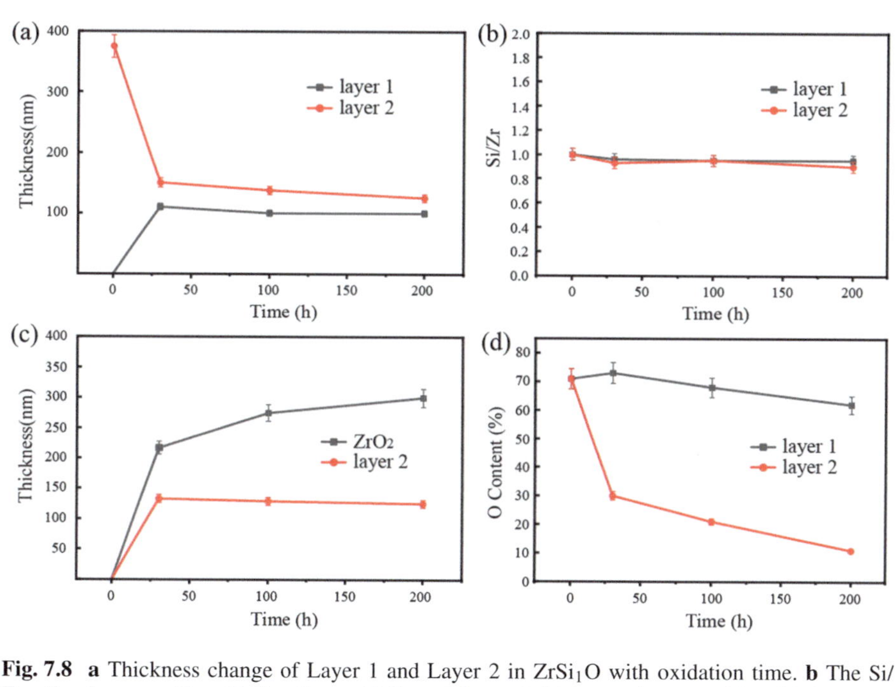

Fig. 7.8 **a** Thickness change of Layer 1 and Layer 2 in $ZrSi_1O$ with oxidation time. **b** The Si/Zr ratio changes with oxidation time. **c** ZrO_2 and Layer 2 thickness changes with oxidation time. **d** Oxygen content changes for Layer 1 and Layer 2

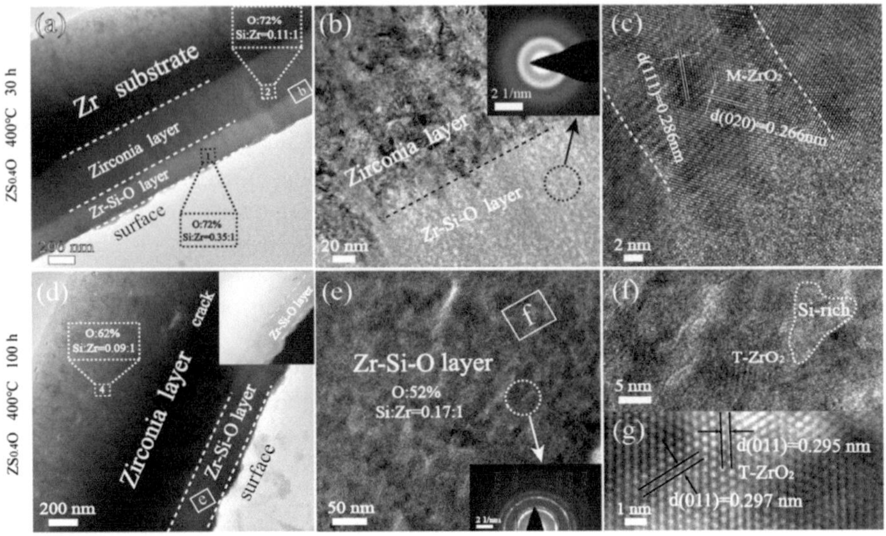

Fig. 7.9 TEM images of the cross-sectional information for $ZS_{0.4}O$-coated Zr alloy after oxidation at 400 °C for 30 h (**a–c**) and 100 h (**d–f**). **a** The bright-field image with EDS analysis after 30 h oxidation. The formation of zirconia layer was attributed to the poor barrier property of $ZS_{0.4}O$. **b** Enlarged image from the marked rectangular box. The insert SAED with halo ring demonstrated that the $ZS_{0.4}O$ layer maintained amorphous structure. **c** HRTEM image of the zirconia/$ZS_{0.4}O$ interface. The formed zirconia is in columnar structure. **d** The bright-field image with EDS analysis after 100 h oxidation. The insert HAADF demonstrated a better contrast for $ZS_{0.4}O$ layer. The protected Zr has totally transformed into zirconia. **e** Enlarged image from the marked rectangular box. The EDS and SAED were shown as insert. **f** HRTEM images

energy, and both silicon and oxygen are not suitable for diffusion. In addition, ZSO with more silicon content has the characteristics of close atomic packing, and the external oxygen diffusion was also effectively prevented. For $ZrSi_{1.8}O$, the element diffusion through the ZSO/Zr-4 interface has suppressed.

Figure 7.10a, b shows the cross-sectional images of $ZrSi_{1.8}O$ after 400 °C/200 h oxidation. $ZrSi_{1.8}O$ was quite stable under oxidation. Not only Si does not diffuse inward, but also the interface reaction becomes difficult. The oxygen-barrier ability was also improved. No ZrO_2 was identified beneath $ZrSi_{1.8}O$ coating. Figure 7.10c, d is the thickness change of Layer 2 and ZrO_2 after 400 °C/200 h oxidation with different ZSO coating. Both Layer 2 and ZrO_2 became thinner with the increased silicon content. The synchronized thickness change of Layer 2 and ZrO_2 again confirmed the interface reaction of ZSO/Zr. The interface reaction of $ZrSi_{1.8}O$/Zr was suppressed. In this case, we further investigated the oxygen-barrier property of ZSO with higher silicon content. The content of solid solution oxygen in Zr alloy at 100 nm below the coating gradually decreased with the increase of silicon after 100 h oxidation. A small amount of silicon diffuses into the zirconium, but remains in much the same content.

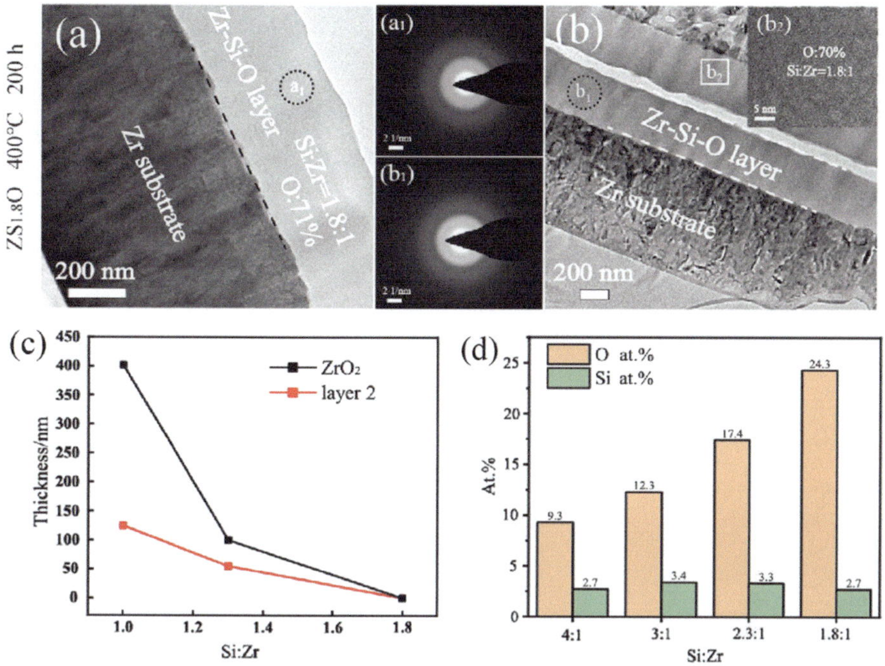

Fig. 7.10 **a** As-deposited $ZrSi_{1.8}O$ layer with EDS analysis. The SAED pattern was shown. **b** The $ZrSi_{1.8}O$ layer after 200 h oxidation with EDS analysis. The SAED pattern was shown. **c** The thickness change of Layer 2 and ZrO_2 of ZSO layers at 400 °C/200 h oxidation with Si:Zr ratio. **d** The content of desolved oxygen in the substrate at 100 nm below the coating with different Si content after 100 h oxidation at 400 °C

The structure evolution of two ZSO coatings with high silicon content (Si/Zr ratio of 3:1 for $ZrSi_3O$ and 4:1 for $ZrSi_4O$) was investigated at 320 °C, 15 MPa in HHW with 0.01 mol/L LiOH solution. Figure 7.11a, b shows the microstructure evolution of $ZrSi_3O$ after 20 h oxidation. ZSO layer does not dissolve in aqueous environment and maintained amorphous state. The SAED patterns indicate that $ZrSi_3O$ maintained in amorphous state. For $ZrSi_4O$, the surface structure of ZSO has undergone significant changes after 5 h oxidation, as shown in Fig. 7.11c, d. The enlarged image demonstrates that the surface layer consists of fine nanoparticles and surrounding amorphous phase. EDS result shows that a large amount of silicon was lost in this layer, with the Zr/Si ratio reduced to 1:1. After 20 h oxidation, the entire amorphous $ZrSi_4O$ film has transformed into crystalline state with a loose structure (Fig. 7.11e). The crystalline phase in the coating is a single-phase tetragonal zirconia. Some interconnected nanotubes were found in this loose structure, and water molecules entered the coating along the nanopores. Simulation shows that water molecules (with a diameter of about 0.3 nm) can diffuse in nanotubes with a diameter as small as about 1 nm. The Zr/Si ratio in the remaining layer is approximately 1:1. Moreover, as shown in the inserted HAADF image, the underneath Zr-4 was oxidized with the

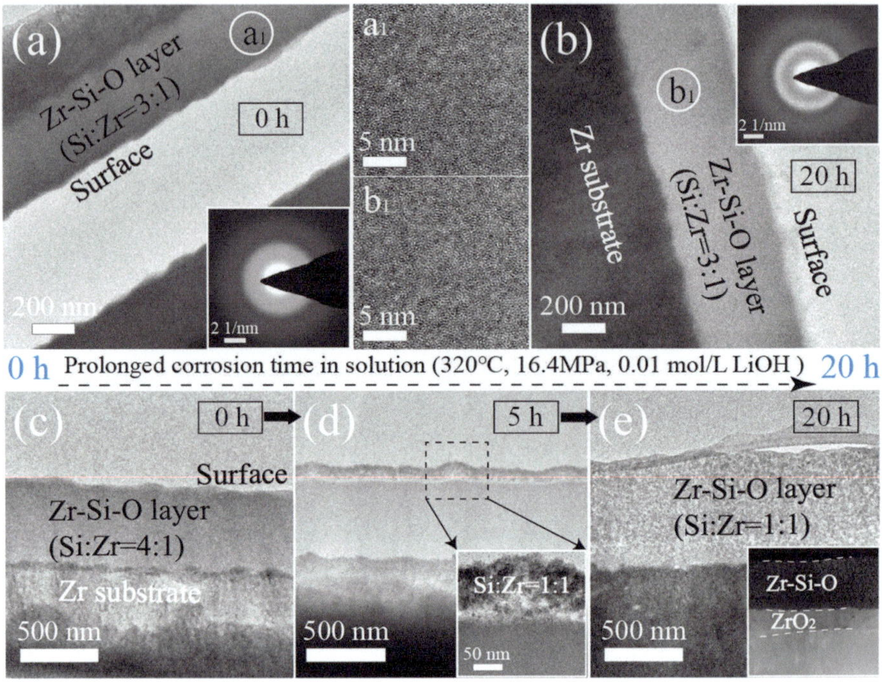

Fig. 7.11 Cross-section TEM images of $ZrSi_3O$ (**a**, **b**) and ZrSi4O (**c**, **d**) after hydrothermal oxidation. **a** The as-deposited $ZrSi_3O$ layer with insert SAED image. The HRTEM image demonstrated the amorphous structure. **b** The $ZrSi_3O$ layer after 20 h hydrothermal oxidation at 320 °C. **c** The as-deposited $ZrSi_4O$ layer. **d** The $ZrSi_4O$ layer after 5 h hydrothermal oxidation at 320 °C. **e** The $ZrSi_4O$ layer after 20 h hydrothermal oxidation at 320 °C. The insert HAADF image demonstrated that the protected Zr has been oxidized

formation of ZrO_2. The local dissolution of silicon was obvious for ZSO with a Zr/Si ratio of 4:1. The Ab initio simulation of the surface reaction between water and oxides reveals the strong exothermic dissociation of water molecules. The simulation revealed the spontaneous dissociation and formation of surface hydroxides. This process is common for many metal oxides that react with water [27]. With high silicon content, it is easier to react with water to form Si–OH bonds and then dissolve in aqueous environment, leading to crystallization of zirconia and initiation of cracks (Fig. 7.12). However, under proper reconstruction, $ZrSi_3O$ coating is chemically stable in HHW (Fig. 7.13).

The bonding conditions of three ZSO coatings under air oxidation and HHW water environments were characterized. XPS results in Fig. 7.14 show that there is an obvious positive shift for O k binding energy (BE) with the increase of silicon content in ZSO. The energy position of $ZrSi_4O$ is close to that of amorphous silicon oxide. For $ZrSi_{1.8}O$ and $ZrSi_3O$, the O k after oxidation in air is not much different from that after oxidation in HHW water. The bonding structure of these two coatings

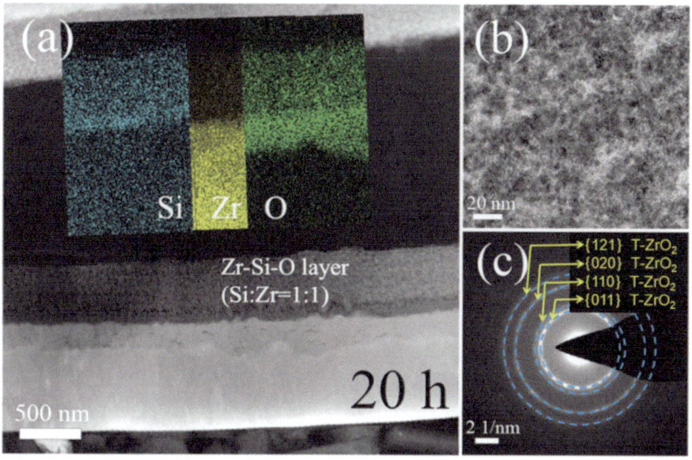

Fig. 7.12 Cross-section TEM of $ZrSi_4O$ after hydrothermal oxidation. Water can readily diffuse through the oxide film with nanopipes or nanopores

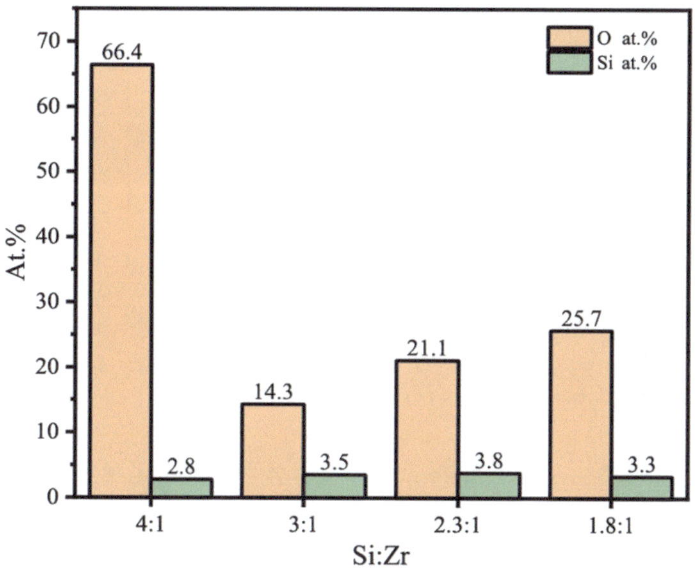

Fig. 7.13 Content of solid solution oxygen in the substrate at 100 nm below the coating with different Si contents after 100 h hydrothermal oxidation

did not change or destroy in HHW water. The O k in ZSO coatings can be divided into two peaks, and the corresponding peak energy is between zirconia and silicon oxide, which confirms that there are two special bonding states of Zr–O–Si inside, one of which has higher silicon content and the other has a lower silicon content.

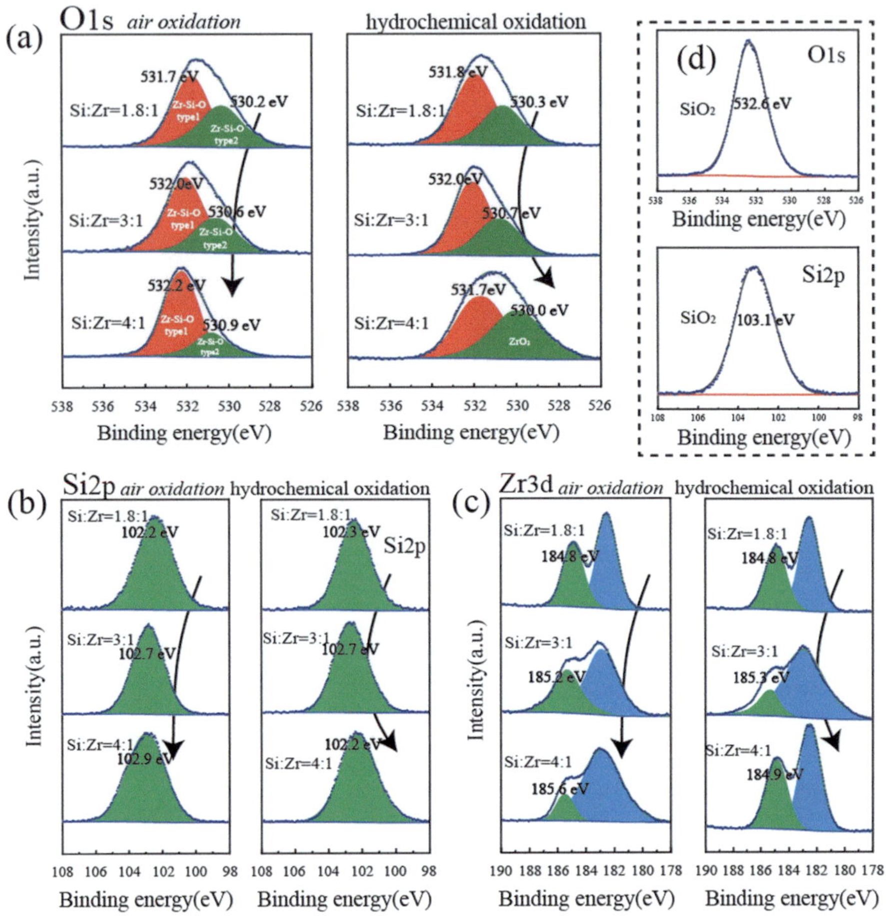

Fig. 7.14 O 1s (**a**), Si 2p (**b**), and Zr 3d (**c**) results after air oxidation and hydrochemical oxidation for 20 h. (**d**) The XPS results of SiO_2 was shown as reference

The spectra are relatively simple for Si 2p and Zr 3d, possibly because the change of bonding state is less sensitive to silicon and zirconium. For $ZrSi_4O$, the bonding state is quite different under HHW water. With the increase of silicon content, the specific atomic bond ratio changes. Zr–O–Si with high BE energy in $ZrSi_4O$ (which is close to the peak of amorphous silicon oxide) disappeared after high-temperature water corrosion. At the same time, a low-energy oxygen BE value is added, which is very close to the peak position in zirconia, corresponding to the crystalline phase in Fig. 7.11e. Meanwhile, the peak strength of Zr 3d also increased significantly, which confirmed the decrease of surface Si content.

On the basis of fully combining theoretical calculation and experimental phenomena, we optimized the best ZSO coating from the perspective of two interfacial reactions. The weight gain of $ZrSi_3O$ protected by the coating was compared

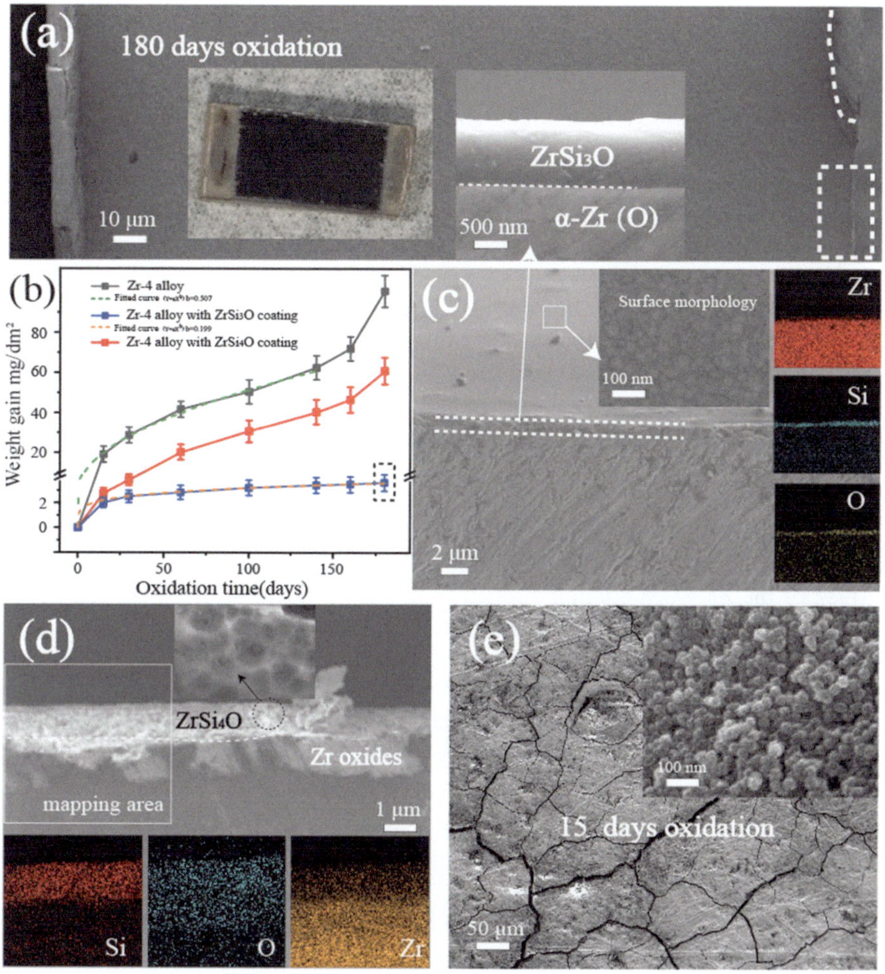

Fig. 7.15 **a**, **b** Cross-sectional and surficial morphology of $ZrSi_3O$ coating after oxidation for 180 days in HHW environment. **c** The weight gain of coated and uncoated Zr-4 alloy after oxidation for 180 days in HHW environment. **d**, **e** The cross-sectional and surface microstructure of $ZrSi_4O$ coating after oxidation for 15 days in HHW environment

to that of the uncoated Zr-4 alloy in Fig. 7.15. Breakaway oxidation occurs after 180 days for the uncoated alloy, and the alloy protected by the coating exhibits stable and extremely low oxidation kinetics, with flat and dense surfaces and stable interfacial bonding. The service performance of $ZrSi_4O$ coating on alloys was also characterized as comparison. After 15 days of corrosion in high-temperature water, the coating structure changed greatly. The dissolution of silicon and the nucleation and growth of zirconia led to more holes in the coating. The structure became extremely loose, and a large number of macroscopic cracks were observed on the surface. Due to the reactive dissolution of silicon, the $ZrSi_4O$ has completely failed under this aqueous environment. Therefore, the silicon content in ZSO is very important for the protection of high-temperature and high-pressure water environment.

The Zr–O–Si coating with a silicon-zirconium atomic ratio of 3:1 has been demonstrated to be stable and provide effective corrosion protection in a high-temperature and high-pressure LiOH solution (320 °C, 16 MPa, 0.01 mol/L), as concluded from the previous discussion. To investigate the oxidation resistance of this coating when applied to zirconium alloys, we deposited a 3-μm-thick Zr–O–Si coating with a silicon-zirconium atomic ratio of 3:1 on Zr-4 alloy. Subsequently, the oxidation resistance was examined at temperatures ranging from 400 °C to 1000 °C in ambient air. Figure 7.16 presents SEM cross-sectional and planar images of the as-deposited state and after oxidation at 400 °C for 100 h in air for the Zr-4 alloy coated with a Zr–O–Si coating (silicon-zirconium atomic ratio of 3:1). In Fig. 7.16a, the planar morphology reveals an intact surface without any cracks or defects on the Zr–O–Si coating deposited on the Zr-4 alloy. Similarly, Fig. 7.16b illustrates that there is no evidence of cracks or defects between the tight bond formed by the coating and substrate in its cross-sectional morphology. After being oxidized at 400 °C for 100 h in air, Fig. 7.16c demonstrates that there are no visible cracks or defects on the surface of the Zr–O–Si coating applied to the Zr-4 alloy. Furthermore, Fig. 7.16d shows that even after undergoing oxidation under these conditions, there are still no signs of cracks or defects observed between both layers based on its cross-sectional image analysis. Additionally, it can be inferred from differences in contrast that this protective coating effectively prevents zirconium alloy oxidation during exposure at elevated temperatures up to approximately 400 °C.

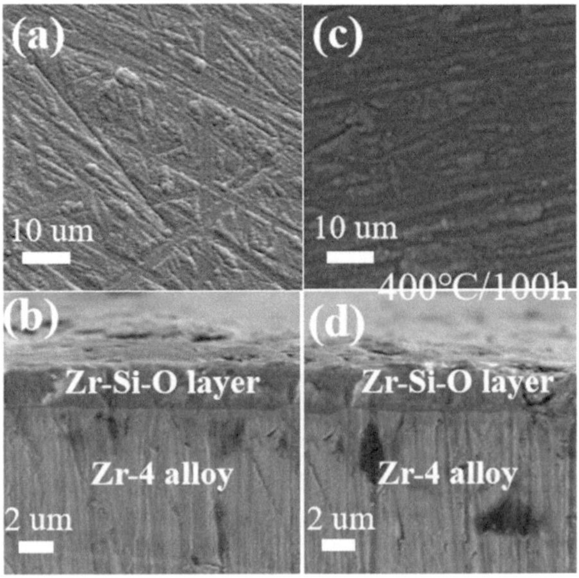

Fig. 7.16 SEM cross-sectional and planar images of Zr-4 alloy protected by a Zr–O–Si coating (with a silicon-zirconium atomic ratio of 3:1) in the as-deposited state and after oxidation at 400 °C for 100 h in ambient air: **a** planar image of the as-deposited Zr–O–Si coating, **b** cross-sectional image of the as-deposited Zr–O–Si coating, **c** planar image of the coating after oxidation at 400 °C for 100 h, **d** cross-sectional image of the coating after oxidation at 400 °C for 100 h

Subsequently, the oxidation temperature was elevated to 600 °C. Figure 7.17 exhibits the SEM cross-sectional and planar images of the Zr-4 alloy coated with a Zr–O–Si protective layer (with a silicon-zirconium atomic ratio of 3:1) after undergoing oxidation at various time intervals at 600 °C in ambient air. Analysis of the planar images following different durations of oxidation at 600 °C reveals the emergence of microcracks on the coating surface. The presence of these cracks potentially facilitates oxygen infiltration, establishing localized regions for direct contact between oxygen and the Zr-4 alloy substrate, thereby expediting alloy oxidation.

The cross-sectional image of the Zr-4 alloy protected by a Zr–O–Si coating after oxidation at 600 °C for 5 h in ambient air is presented in Fig. 7.17b. It can be observed that, following oxidation for 5 h at 600 °C, the Zr-4 alloy with the protective coating exhibits an oxide layer depth of approximately 3.2 μm. Upon extending the oxidation time to 20 h, the thickness of the oxide layer on the coated Zr-4 alloy increases to around 9.3 μm. Furthermore, with a further extension of the oxidation time to 80 h, significant transverse cracks appear on the oxidized Zr-4 alloy substrate along with an increase in oxide layer thickness to approximately 35 μm.

The EDS elemental distribution map of the Zr–O–Si coating (with a silicon-zirconium atomic ratio of 3:1) after oxidation at 600 °C for 20 h in ambient air is

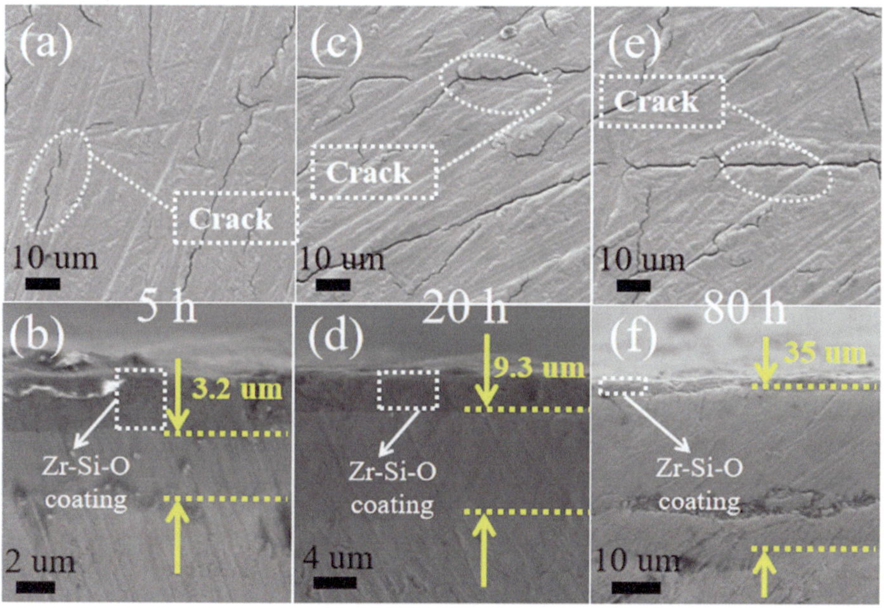

Fig. 7.17 SEM cross-sectional and planar images of the Zr–O–Si coating (with a silicon-zirconium atomic ratio of 3:1) after oxidation at different times at 600 °C in ambient air: **a** SEM planar image after 5 h of oxidation at 600 °C. **b** SEM cross-sectional image after 5 h of oxidation at 600 °C. **c** SEM planar image after 20 h of oxidation at 600 °C. **d** SEM cross-sectional image after 20 h of oxidation at 600 °C. **e** SEM planar image after 80 h of oxidation at 600 °C. **f** SEM cross-sectional image after 80 h of oxidation at 600 °C

Fig. 7.18 SEM cross-sectional elemental distribution map of the Zr–O–Si coating (with a silicon-zirconium atomic ratio of 3:1) after oxidation at 600 °C for 20 h in ambient air

presented in Fig. 7.18. It can be observed that even after undergoing a prolonged exposure to high-temperature and atmospheric conditions, the elements within the coating maintain a uniform distribution, while the coating itself remains intact without any noticeable cracks or defects such as pores. Furthermore, by examining the elemental distribution map, it becomes evident that there is clear evidence of oxidation on the Zr-4 alloy substrate.

Based on previous research and discussions, it has been established that a Zr–O–Si coating with a silicon-zirconium atomic ratio of 3:1 effectively protects the Zr-4 alloy during oxidation in air at both 400 and 600 °C. Consequently, we will now proceed to investigate the oxidation behavior at an elevated temperature of 800 °C. Figure 7.19 presents SEM cross-sectional and planar morphology images of the zirconium alloy protected by the Zr–O–Si coating (with a silicon-zirconium atomic ratio of 3:1) after undergoing oxidation for varying durations at 800 °C in ambient air. Notably, micro-scale cracks are observed on the surface of the coating following oxidation at this higher temperature. Comparatively, these cracks appear wider and deeper than those observed after oxidation in ambient air at 600 °C; however, there is no apparent increase in crack density.

The SEM cross-section of Zr-4 alloy, protected by a Zr–O–Si coating and subjected to oxidation at 800 °C for 1 h, is depicted in Fig. 7.19. It can be observed that the coating exhibits an approximate thickness of 3 μm and remains firmly adhered to the Zr-4 alloy substrate even after exposure to high temperatures. Under these conditions, the coated Zr-4 alloy experiences an oxidation depth of about 10 μm. Upon extending the air oxidation time to 2 h at 800 °C, the uncoated Zr-4 alloy demonstrates a significant increase in oxide film thickness up to 57 μm. This illustration clearly demonstrates the phenomenon of thickening oxide film under protection from the Zr–O–Si coating, which reaches approximately15 μm on the surface of coated Zr-4 alloy. Further prolonging air oxidation time to 8 h at 800 °C (Fig. 3.4f) reveals a

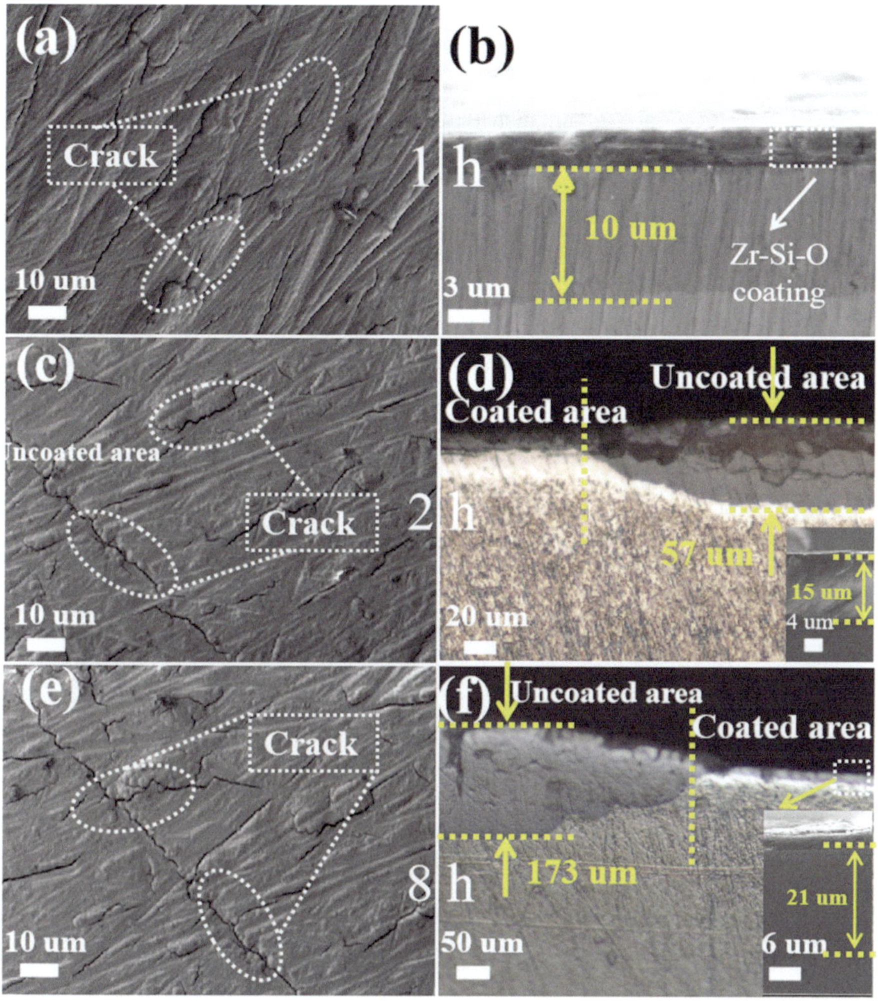

Fig. 7.19 SEM cross-section and plane diagram of Zr–O–Si coating with 3:1 Si-Zr atomic ratio after oxidation at 800 °C for different times in air environment: **a** SEM diagram of Zr–O–Si coating oxidized at 800 °C for 1 h, **b** SEM cross-section diagram of Zr–O–Si coating oxidized at 800 °C for 1 h, **c** SEM diagram of Zr–O–Si coating oxidized at 800 °C for 2 h, **d** SEM cross-section diagram of Zr–O–Si coating oxidized at 800 °C for 2 h, **e** SEM diagram of Zr–O–Si coating oxidized at 800 °C for 8 h, **f** SEM cross-section diagram of Zr–O–Si coating oxidized at 800 °C for 8 h

cross-sectional view depicting the interface between the protective Zr–O–Si coating and bare zirconium alloy region. Notably, only a thin layer of oxidation is observed on zirconium alloy protected by this coating while extensive oxidization process takes place resulting in nearly 200μm thickness on bare zirconium alloy area. The illustration provides additional evidence regarding thickening behavior of oxide film under protection from the Zr–O–Si coating, reaching approximately 21 μm on the coated Zr-4 alloy. Therefore, the Zr–O–Si coating with a silicon-zirconium atomic ratio of 3:1 can still effectively protect the Zr-4 alloys at temperatures as high as 800 °C.

The EDS surface scan element distribution of the cross-section of Zr–O–Si coating with a silicon-zirconium atomic ratio of 3:1 after oxidation at 800 °C/8 h in an air environment is presented in Fig. 7.20. From the SEM cross-section analysis of the Zr-4 alloy without coating protection after oxidation at 800 °C/8 h in air (Fig. 7.20a), numerous microcracks can be observed, leading to a loose structure that fails to

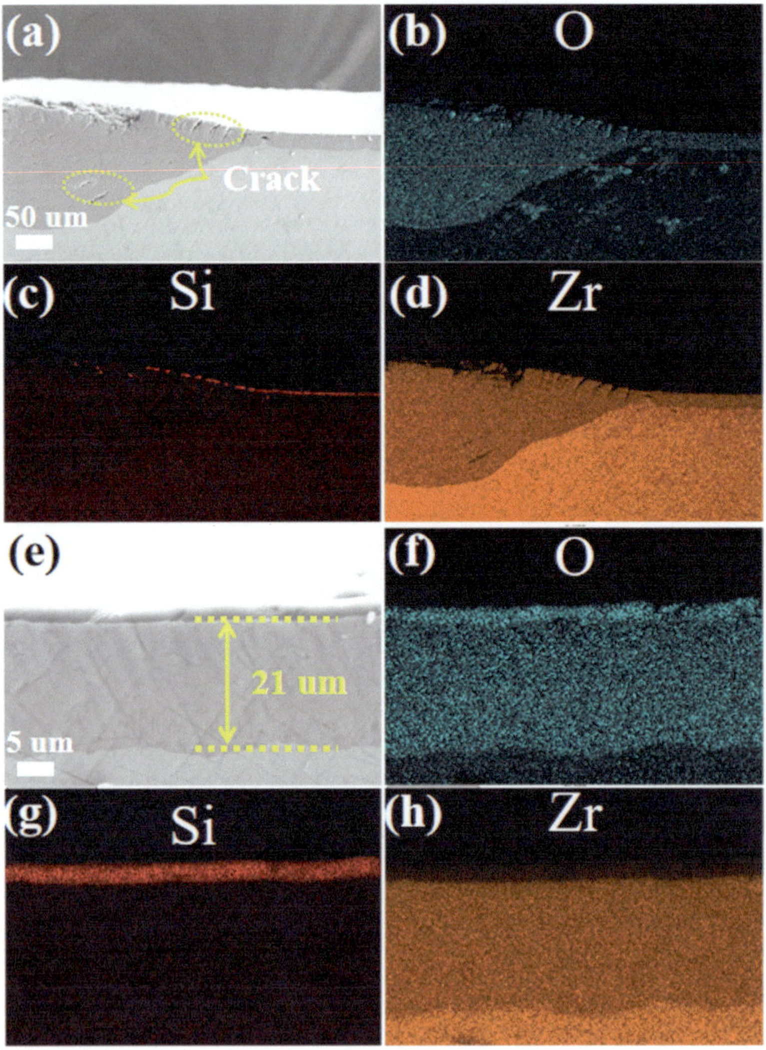

Fig. 7.20 Distribution of EDS elements on SEM cross-section of Zr–O–Si coating with 3:1 Si-Zr atomic ratio after oxidation at 800 °C at different time in air environment

impede oxygen diffusion into the inner part of the alloy. Figure 7.20e displays the cross-sectional topography of the Zr-4 alloy with Zr–O–Si coating protection after oxidation at 800 °C/8 h in an air environment. It is evident that although there exists a certain thickness of oxide layer formation where coating protection is present, no cracks or significant defects are observed, and the structural organization remains compact and intact. The element distribution analysis across the interface position between Zr–O–Si coating and bare zirconium alloy (Fig. 7.20b, c, d) intuitively reveals variations in oxidation thickness between areas with and without coating protection, further demonstrating successful incorporation of our prepared elements—Zr, Si, and O. Based on the elemental distribution maps of the alloy cross-section with Zr–O–Si coating (Fig. 7.20f, g, h), it can be observed that there is no significant inward diffusion of Zr and Si elements. Combined with the composition analysis conducted through EDS point scanning, it is determined that the Zr-4 alloy protected by the coating undergoes oxidation to form pure ZrO_2.

The variation curve of oxide film thickness for bare Zr-4 alloy and Zr–O–Si-coated Zr-4 alloy with increasing oxidation time at temperatures of 400 and 800 °C is presented in Fig. 7.21. It can be observed that both the bare Zr-4 alloy and the coated Zr-4 alloy exhibit a rapid initial growth period of the oxide film during oxidation at the same temperature, followed by a gradual decrease in growth rate as oxidation time progresses. Generally, the growth rate of the uncoated Zr-4 alloy oxide film is higher compared to that of the coated counterpart, indicating that the presence of a protective Zr–O–Si coating slows down oxide film growth and provides certain protective effects.

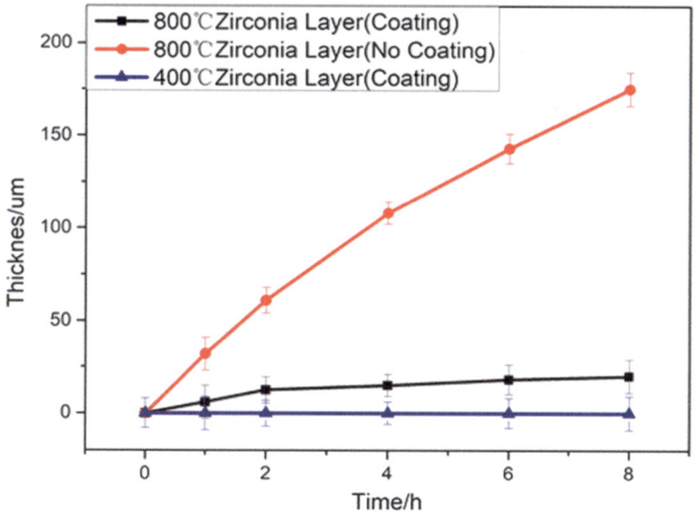

Fig. 7.21 Thickening curves of oxide film of Zr-4 alloy protected by bare Zr-4 alloy and Zr–O–Si coating with 3:1 zircon atomic ratio at different temperatures

7.4 Conclusion

The amorphous ZSO coatings were designed for the protection of Zr alloys in nuclear working conditions. Oxidation-induced elemental diffusion and interfacial reactions (Zr/ZSO, ZSO/H_2O) were determined by the bonding status and the chemical difference in ZSO. $ZrSi_3O$ was optimized and could be regarded as Zr-modified amorphous silicon oxide. On the one hand, ensuring the dominance of silicon in ZSO is the basis of oxidation resistance, which guaranteed the atomic structure dominated by silica tetrahedron and suppressed the inner diffusion of Si and O from ZSO to Zr. On the other hand, the addition of Zr is reflected in inhibiting the reaction of Si and high-temperature water. The introduction of Zr ions caused the change of amorphous bonding structure from –Si–O–Si– to –Zr–O–Si– and regulated the electronic structure of the silica tetrahedron. The adsorption energy of hydroxyl on silicon atoms was increased (combined catalysis). Thus the reaction dissolution of silicon in high-temperature and high-pressure water was inhibited.

When the Zr is dominant in ZSO, both Si and O diffusion occurred at the interface of ZSO/Zr-4 alloy. The inward diffusion of silicon from ZSO into zirconium alloy is an oxidation-induced process. The surface ZSO layer gradually crystallized into tetragonal zirconia due to the decrease of silicon content, losing its protective ability. With increased Si content, the inward diffusion of silicon was mitigated. The ZSO/Zr interface reaction occurred as a redox reaction due to the unbalanced chemical potential. Zr-4 alloy in contact with ZSO coating transformed into T-ZrO_2 layer and the ZSO coating evolved into a two-layer structure. With the further increase of silicon content, the atomic arrangement of amorphous ZSO transformed from Zr-dominant to Si-dominant and becomes denser. The atomic packing status changed as well as the element diffusion behavior. The element diffusion and redox reaction are obviously inhibited. However, more silicon in ZSO can cause corrosive damage in high-temperature and high-pressure water environments. A trade-off of "diffusion barrier" and "hydrochemical stability" was achieved in the optimized $ZrSi_3O$ coating in this work, which has the potential to be used in light-water reactors.

References

1. Duan Z, Yang H, Satoh Y et al (2017) Current status of materials development of nuclear fuel cladding tubes for light water reactors. Nucl Eng Des 316:131–150
2. Allen T, Busby J, Meyer M et al (2010) Materials challenges for nuclear systems. Mater Today 13(12):14–23
3. Hallstadius L, Johnson S, Lahoda E (2012) Cladding for high performance fuel. Prog Nucl Energy 57:71–76
4. Turque I, Chosson R, Le Saux M, Brachet JC, Vandenberghe V, Crépin J, Gourgues-Lorenzon AF (2018) Mechanical behavior at high temperatures of highly oxygen- or hydrogen-enriched α and prior-β phases of zirconium alloys, zirconium in the nuclear industry: 18th international symposium, ASTM STP1597, ASTM international, west conshohocken, PA, pp 240–280
5. Bell BDC, Murphy ST, Burr PA et al (2016) The influence of alloying elements on the corrosion of Zr alloys. Corros Sci 105:36–43

6. Kim HG, Kim IH, Park JY et al (2015) Application of coating technology on zirconium-based alloy to decrease high-temperature oxidation. Zirconium in the Nuclear Industry: 17th Volume. ASTM International
7. Park JH, Kim HG, Park J et al (2015) High temperature steam-oxidation behavior of arc ion plated Cr coatings for accident tolerant fuel claddings. Surf Coat Technol 280:256–259
8. Cao G, Chen C, Xu H, Ban J, Liu F, Yuan G, Lei H, Su Y, Hu J (2023) Chemical regulation in the bonding reconstruction stage of amorphous Zr–Si oxidation and the resultant phase selection. Mater Des 233:112291
9. Pei S, Yang F, Feng N, Hu J, Shao G, Yuan G, Cao G (2022) On the thermal stability and oxidation resistance of Zr/X (Cr, Ni, Si) multilayer structure. Surf Coat Technol 440:128500
10. Wu K, Yao H, Cheng X, Hu J, Cao G, Yuan G (2020) Oxidation behavior and chemical evolution of architecturally arranged Zr/Si multilayer at high temperature. Surface and Coatings Techn 126205
11. Lim KR, Park JM, Kim SJ et al (2013) Enhancement of oxidation resistance of the supercooled liquid in Cu-Zr-based metallic glass by forming an amorphous oxide layer with high thermal stability. Corros Sci 66:1–4
12. Liang B, Yang ZH, Jia DC et al (2016) Amorphous silicoboron carbonitride monoliths resistant to flowing air up to 1800 °C. Corros Sci 109:162–173
13. Abdollahi Boraei SB, Esmaeili Bidhendib M, Afzali D (2017) Preparation of SiO_2/ZrO_2 ceramic nanocomposite coating on Aluminum alloys as metallic part of the photovoltaic cells and study its corrosion behavior. Environ Energy Econom Res 1(2):231–238
14. Chen L, Li J, Zhang Y et al (2015) Effects of alloyed Si on the autoclave corrosion performance and periodic corrosion kinetics in Zr–Sn–Nb–Fe–O alloys. Corros Sci 100:651–662
15. Yeom H, Maier B, Mariani R, Bai D, Sridharan K, Fronek S, Xu P (2017) Magnetron sputter deposition of zirconium-silicide coating for mitigating high temperature oxidation of zirconium-alloy. Surf Coat Technol 316:30–38
16. Yeom H, Lockhart C, Mariani R, Peng C, Michael S (2018) Evaluation of steam corrosion and water quenching behavior of zirconium-silicide coated LWR fuel claddings. J Nuclear Mater Mater Aspects Fission Fusion 499:256–267
17. Xu H, Xiao W, Wang Z, Hu J, Shao G (2021) Self-consistent assessment of Li+ ion cathodes: theory vs. experiments. J Energy Chem 59:229–241
18. Dasmahapatra A, Meletis E, Kroll P (2017) First principles modeling and simulation of Zr–Si–B–C–N ceramics: developing hard and oxidation resistant coatings. Acta Mater 125:246–254
19. Farabaugh EN, Feldman A, Sun J et al (1987) Examination of thin films in the ZrO_2-SiO_2 system by transmission electron microscopy and x-ray diffraction techniques. J Vac Sci Technol, A: Vac, Surf Films 5(4):1671–1674
20. Luo WD, Yang B, Chen GL (2011) Effect of Ar+ ion irradiation on the microstructure and properties of Zr-Cu-Fe-Al bulk metallic glass. Scr Mater 64(7):625–628
21. Jiang N (2009) Electron energy-loss near-edge structure of zircon. Mater Res Soc Symp Proc 1124:3–9
22. Jiang N, Spence J (2009) Interpretation of O K-edge EELS in zircon using a structural variation approach. Ultramicroscopy 110(1):89–93
23. Batson PE (1993) Silicon $L_{2,3}$ near-edge fine structure in confined volumes. Ultramicroscopy 50(1):1–12
24. Nonlinear composition dependence of x-ray photoelectron spectroscopy and Auger electron spectroscopy features in plasma-deposited zirconium silicate alloy thin films
25. Yoshiya M, Adachi H, Tanaka I (1999) Interpretation of Si-$L_{2,3}$ edge electron energy loss near edge structures (elnes) from intergranular glassy film of Si_3N_4 ceramics. J Am Ceram Soc 82(11):3231–3236
26. Kim WJ et al (2002) Corrosion behavior of reaction-bonded silicon carbide ceramics in high-temperature water. J Mmater Sci Lett 21(9):733–735
27. Wen B et al (2023) Water dissociation at the water–rutile TiO_2(110) interface from Ab initio-based deep neural network simulations proceedings of the national academy of sciences - PNAS 120(2):1–e2212250120 (2023)

Chapter 8
Application of High-Entropy Amorphous Oxide Coating in Zirconium Alloys

Abstract High-entropy multicomponent oxide coatings have emerged as a promising material due to the enhanced elemental miscibility and glass forming ability. This chapter designed and fabricated ZrNbCrAlSiO coatings for the protection of zirconium alloys. After surface hypoxic treatment, the formation of (Zr, Nb, Cr, Al)M–O–Si mixed bonds on the surface enhances the structural stability of $(ZNCA)_1S_3$ at subcritical environment. At 1000 °C, a stable high-entropy amorphous nanocrystalline structure was formed in $(ZNCA)_1S_3$, significantly reduces the grain coarsening, alleviates stress induced by high-temperature oxidation, and minimizes surface cracking.

ncbvncbvncbvnc

Keywords Multicomponent oxide coatings · Amorphous stability · High-temperature oxidation · Corrosion resistance · Zirconium alloys

8.1 Introduction

In the 2011 Fukushima nuclear accident, zirconium alloys underwent a rapid oxidation reaction under high-temperature steam conditions, and the resulting hydrogen gas triggered an explosion. Currently, there are two main ways to improve the accident resistance of cladding materials. Long-term development is to design new nuclear reactor cladding materials to replace zirconium alloys, solving the inherent defects of zirconium alloys in high-temperature water vapor environment. FeCrAl alloys and SiC composites can be used as the replacement materials of zirconium alloys used in nuclear reactor cladding [1–4]. The key to short-term development is to ensure that the accident tolerance of zirconium alloys is improved without affecting their normal operational stability. The research and development of accident-tolerant fuel (ATF) are mainly based on the surface modification of zirconium alloy [5, 6].

The reported results show that metal Cr, Fe–Cr–Al, MAX phase, etc., as zirconium alloy protective coatings are deficient in some aspects, such as coating-matrix interface diffusion induced formation of brittle phases, which damage the structure

J. Hu and G. Cao, *Zirconium Alloy Coatings*,
https://doi.org/10.1007/978-981-96-5028-6_8

of the zirconium alloy itself. The diffusion of the elements in oxidation and corrosion process of the protective coating will lead to the emergence of microporous, which in turn affects the coating's protective effect [7–10]. Therefore, the surface modification of zirconium alloys for nuclear reactor cladding still has a long way to go.

The antioxidant property of zirconium alloy itself is affected by the growth of zirconium oxide on the surface. The inward growth of zirconium oxide will generate internal stresses and creaks, which will accelerate the oxidation of zirconium alloys. The volume change induced by phase transformation is the main reason for the failure of the protective zirconia layer [11–14]. Musil et al. inhibited the transformation of the T-ZrO_2 to the M-ZrO_2 phase by doping a certain amount of silicon into ZrO_2, avoiding the appearance of the phase transformation defects, and then improve the protective properties of zirconia [15]. Si-based materials have very low-oxygen solid solubility and oxygen diffusion coefficients, and SiO_2 exists mostly in an amorphous form in the multimaterial system [16, 17]. The amorphous SiO_2 has a certain mobility to repair the microcracks and micropores generated in the oxidation process, which makes the whole multimaterial system have a certain self-repairing function. However, SiO_2-based protective coatings were unstable in high-temperature and high-pressure water. Usui. T et al. studied the application of SiC on zirconium alloys. Carbide coating materials have better structural stability in the irradiation environment. However, SiC coating in the high-temperature and high-pressure water environment will be oxidized to form SiO_2 oxygen-barrier layer and react with high-temperature and high-pressure water vapor, dissolved in the corrosive liquid [18, 19].

In the previous section, it was demonstrated that Zr–O–Si coatings can modulate the network structure of the silica-oxygen tetrahedra and improve the hydrochemical stability. However, it does not have a protective effect at higher temperatures. T-ZrO_2 grains crystallized after the oxidation of the Zr–O–Si coating is a fast oxygen ion diffusion channel, high-temperature oxidation grains coarsened, resulting in the coating to lose the nanostrengthening effect, easy to form extended cracks. The coating loses the nanostrengthening effect and is easy to form extended cracks. Multicomponent structure of high mixing entropy is conducive to enhance the mutual solubility of the elements, slow down the diffusion effect, inhibit the formation of the crystalline phase, and stabilize the amorphous system. The crystallization phenomenon during high-temperature oxidation of oxide coatings can be suppressed by adding components, and a kind of multioxide coating material with a stable structure at high temperature in accordance with the endowed characteristics of zirconium alloys, interfacial compatibility, good oxygen-barrier effect, and mechanical properties can be prepared.

When designing zirconium alloy multicomponent oxide protective coatings for nuclear reactors, the selection of Zr-containing coatings ensures good interfacial bonding between the coatings and the zirconium-based alloys. Metallic elements such as Cr, Al, Nb, and Zr, all of which have low thermal neutron absorption cross-sections, can be utilized in the design of zirconium alloy protective coatings for nuclear reactor claddings. Cr and Nb are two typical refractory metals often used to improve the high-temperature mechanical properties of alloys or coating materials. Cr and Al are important constituent elements to improve the corrosion resistance of metals or alloys, and the incorporation of these two elements can improve the high-temperature oxidation resistance of the coatings to a certain extent. Oxide ceramic coatings have higher oxidation resistance and irradiation resistance than high-entropy alloy coatings. The high mixing entropy of the multicomponent oxide structure is conducive to enhancing the mutual solubility of the elements, slowing down the diffusion effect between the elements, inhibiting the formation of crystalline phases, and stabilizing the amorphous system. Based on this, ZrNbCrAlSiO coatings are prepared in this chapter to investigate their structural stability after high-temperature corrosion and oxidation and their protective effect when actually used as protective coatings for Zr-4 alloys.

8.2 Experimental Method

The ZrNbCrAlSiO coatings were deposited using the S500 HiTUS system. The Zr, Cr, Al, and Nb elemental targets were separately cut and respliced and combined to form a complete sputtering target, which is mounted onto the target position connected to a DC power supply. The Si target was mounted on the target position connected to the RF power supply. Then oxygen was passed into the chamber to prepare the ZrNbCrAlSiO coating by reactive sputtering. For the convenience of describing the different compositions of the coatings in the article, the coatings with ZrNbCrAl:Si = 1:n are denoted as $(ZNCA)_1S_n$.

The surface analysis was performed using X-ray photoelectron spectroscopy (XPS, AXIS Ultra DLD). In this experiment, to remove surface hydrocarbon pollutants, a 200eV argon ion polishing device was used to gently clean the sample surface. The surface and cross-sectional morphology of the coating were studied using a scanning electron microscope (SEM, Zeiss Sigma 500). The microstructure characteristics of the prepared coating were analyzed based on transmission electron microscopy (TEM, FEI Talos F200S) and energy-dispersive X-ray spectroscopy (EDS, X-Max Extreme, energy resolution: Mn-K α and 10 kcps (output) $\leq$ 125 eV). Prepare TEM cross-sectional analysis samples using the Gatan 691 PIPS instrument.

8.3 Study on the High-Temperature Oxidation Behavior of ZrNbCrAlSiO Coating

Figure 8.1 shows the TEM cross-sectional morphology of the prepared $(ZNCA)_1S_{0.09}$ coatings with lower silicon content annealed at 800 °C and 1000 °C in air for 2 h, respectively. At this time, the percentage of silicon atoms in the coating is about 2.92%, which accounts for a relatively small amount of content, equivalent to a small amount of silicon doped in the alloy oxide coating. Figure 8.1a and its inset show that after oxidation the $(ZNCA)_1S_{0.09}$ coating at 800 °C for 2 h in an air environment,

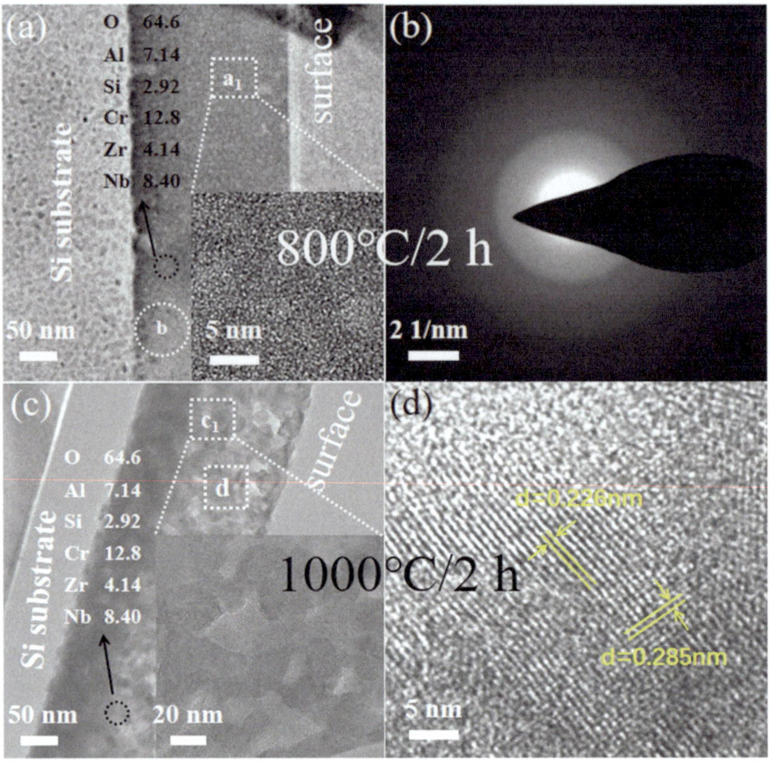

Fig. 8.1 TEM images of $(ZNCA)_1S_{0.09}$ coating after oxidation at different temperatures in air **a** and **b** are cross-sectional morphology and SAED images after oxidation at 800 °C for 2 h respectively **c** and **d** are cross-sectional morphology and high-resolution images after oxidation at 1000 °C for 2 h, respectively

the coating is still firmly bonded to the silicon substrate, and there are no cracks or holes in the coating. Figure 8.1a shows that the coating is still amorphous after oxidation at 800 °C for 2 h. Figure 8.1c shows the TEM cross-section morphology after further increasing the oxidation temperature to 1000 °C for 2 h. Figure 8.1c and the inset (partial enlargement) show that after oxidation at 1000 °C for 2 h, the coating appears to be crystallized, and the original dense structure is destroyed, and many large grains are crystallized from the coating.

Figure 8.2 shows the EDS surface swept element distribution of $(ZNCA)_1S_{0.09}$ coating after oxidation for 2 h at 1000 °C air environment, at this time the coating did not show dramatic element diffusion and loss phenomenon, the elements in the coating are uniformly distributed, but the coating of the lower edge of the silicon wafer was oxidized to a certain thickness, and the oxygen-barrier ability of the multicomponent oxide coating with a lower silicon content is general. In order to further investigate the protective effect of $(ZNCA)_1S_{0.09}$ coating on Zr-4 alloy when it is actually applied on it. $(ZNCA)_1S_{0.09}$ coatings with a thickness of 3 μm were further

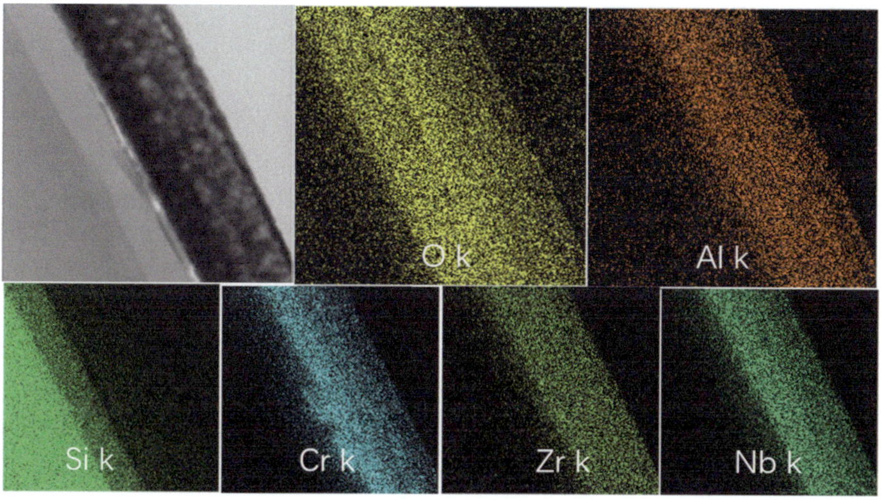

Fig. 8.2 EDS surface sweep element distribution of $(ZNCA)_1S_{0.09}$ coating annealed at 1000 °C in air for 2 h

prepared on Zr-4 alloy and oxidized in an air environment at 800 °C and 1000 °C for 2 h, respectively, and then the thickness of the oxide film with and without the coating protection position was characterized.

Figure 8.3 shows the SEM surface and cross-sectional morphology of the $(ZNCA)_1S_{0.09}$-coated Zr-4 alloy in the as-prepared state and after oxidation at 800 °C and 1000 °C for 2 h in an air environment. Figure 8.3a shows the SEM surface morphology of the prepared state of the coating, which shows that the surface structure is relatively intact (the scratches on the surface are produced by the zirconium alloy during the polishing process), and there are no cracks on the surface. Figure 8.3b shows the SEM cross-section of the coating in the prepared state, which shows that the coating is tightly bonded to the Zr-4 alloy substrate, and there are no defects such as cracks and holes in the cross-section of the coating. Figure 8.3c shows the surface morphology of the Zr-4 alloy protected by the coating after oxidation at 800 °C for 2 h in an air environment, and it can be seen that no microcracks were observed on the surface. Figure 8.3d shows the SEM cross-section of the Zr-4 alloy interface with and without coating protection after oxidation at 800 °C for 2 hours. Without coating protection, the Zr-4 alloy forms an oxide film approximately 60 micrometers thick. In contrast, the (ZNCA)1S0.09-coated Zr-4 alloy exhibits a much thinner oxide layer, with a growth of only about 16 micrometers. At this time the coating played a certain protective effect on Zr-4 alloy. The SEM cross-sectional morphology after oxidation can be observed that there are transverse cracks due to zirconia phase transition at the uncoated position. Figure 8.3e and f is the SEM surface morphology and cross-section morphology after oxidation at 1000 °C for 2 h, and the cross-section morphology shows that the thickness of the oxide film of Zr-4 alloy is basically the same in the position with or without coating protection, and the coating loses its protective effect

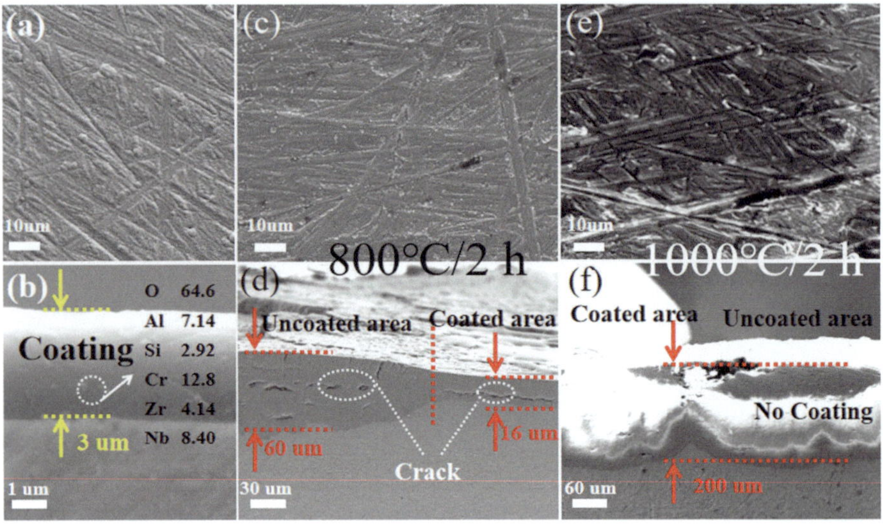

Fig. 8.3 SEM surface and cross-sectional morphology of Zr-4 alloy protected by $(ZNCA)_1S_{0.09}$ coating in the prepared state and after oxidation at 800 °C and 1000 °C for 2 h: **a** and **b** surface and cross-sectional morphology in the prepared state, respectively **c** and **d** surface and cross-sectional morphology after oxidation at 800 °C for 2 h, respectively **e** and **f** surface and cross-sectional morphology after oxidation at 1000 °C for 2 h

at this time. The $(ZNCA)_1S_{0.09}$ coating prepared based on the Zr–O–Si protective coating can play a certain protective effect on Zr-4 alloy when oxidized in the air environment at 800 °C, but when the oxidation temperature rises to 1000 °C, the coating loses its protective effect. The coating failure during oxidation at 1000 °C in air may be attributed to crystallization-induced grain coarsening. The formation of large grains creates intercrystalline defects, which serve as fast diffusion channels for oxygen, thereby accelerating oxidation at the coating/Zr-4 alloy interface.

The results show that $(ZNCA)_1S_{0.09}$ coating has amorphous structure after 2 h of air oxidation at 800°C, which also has a certain protective effect on Zr-4 alloy. After oxidation at 1000°C for 2 h, the coating crystallizes into large grains, and its protective effect on Zr-4 alloy is completely lost. Silicon is an element with low-oxygen diffusion coefficient, and SiO_2 is mostly in an amorphous state in the multifunctional system and can stabilize the amorphous system. Based on this characteristic of silicon, on the basis of $(ZNCA)_1S_{0.09}$ coating prepared in the previous section, the content of silicon element was increased, $(ZNCA)_1S_3$ coating with relatively high silicon content was prepared, and oxidation experiments at different temperatures in the air environment were carried out separately to investigate the structural stability of $(ZNCA)_1S_3$ coating in the high-temperature air environment and the protective performance of Zr-4 alloy.

The $(ZNCA)_1S_3$ coating was subjected to oxidation at various temperatures in an air environment to assess its structural stability and protective performance on Zr-4 alloy. Sectional TEM morphologies of ZrNbCrAlSiO coating with high silicon content rapidly annealed at 800, 900, and 1000 °C for 2 h are shown in Fig. 8.4. From Fig. 8.4a, it can be observed that the microstructure of $(ZNCA)_1S_3$ remains stable after oxidation at 800 °C for 2 h, without any observed defects such as cracks or holes. High-resolution TEM images in Fig. 8.4 (a_1) demonstrate that the coating retains its amorphous state even after oxidation at 800°C for 2 h, indicating a strong ability to maintain amorphous formation. The TEM topographic images in Fig. 8.4b and c depict the cross-section of $(ZNCA)_1S_3$, which were annealed at 900 °C and 1000 °C for 2 h in an air environment, respectively. It is evident that no extensive crystallization phenomenon occurs in $(ZNCA)_1S_3$ after exposure to high-temperature oxidation. Instead, a structure comprising nanocrystals of amorphous coating emerges. Figure 8.5 displays EDS surface scanning results of the region containing this amorphous nanocrystalline structure, revealing an even distribution of elements without segregation or aggregation following high-temperature oxidation. The selected area electron diffraction structure of the nanocrystal structure is depicted in Fig. 8.6, revealing that the material composition of the nanocrystal particles formed after oxidation at 900 °C and 1000 °C for 2 h remains consistent, consisting of both tetragonal and hexagonal phases. Consequently, it can be inferred that the amorphous phase within the amorphous nanocrystal structure primarily comprises amorphous SiO_2, while the nanocrystal structure mainly consists of high-entropy oxide particles in a combination of tetragonal and hexagonal phases. Statistical analysis conducted on a large number of nanocrystals reveals that after oxidation at 900 °C for 2 h,

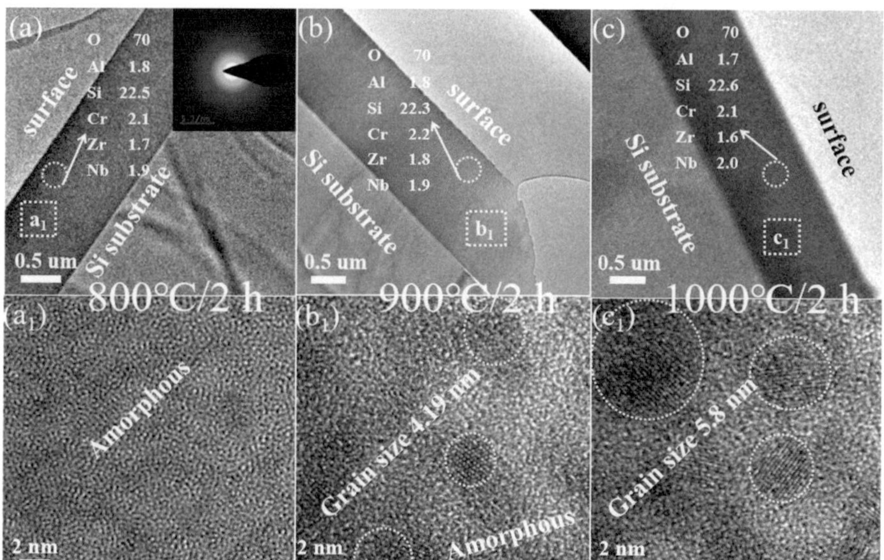

Fig. 8.4 Sectional TEM topography of ZrNbCrAlSiO film/coating with high silicon content rapidly annealed at 800 °C (**a**), 900 °C (**b**), and 1000 °C (**c**) for 2 h, respectively. (a1), (b1), and (c1) are the corresponding TEM high-resolution images

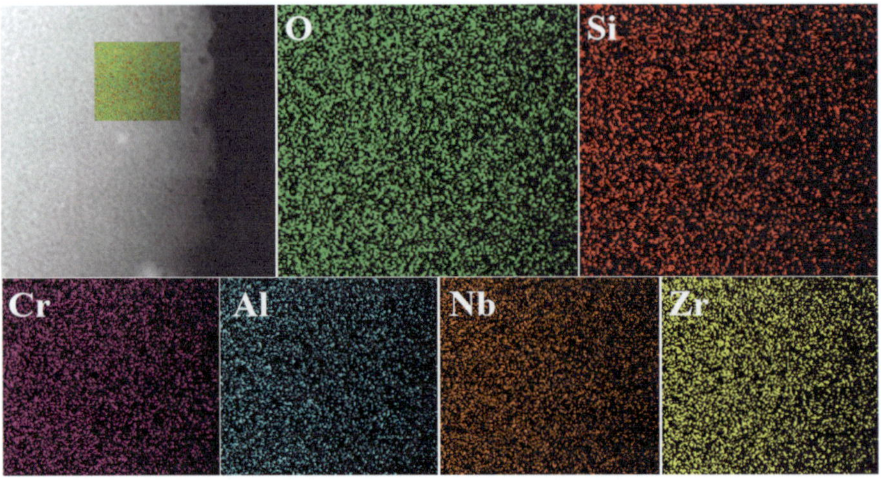

Fig. 8.5 Sectional EDS surface scan of ZrNbCrAlSiO film/coating with high silicon content rapidly annealed at 1000°C for 2 h

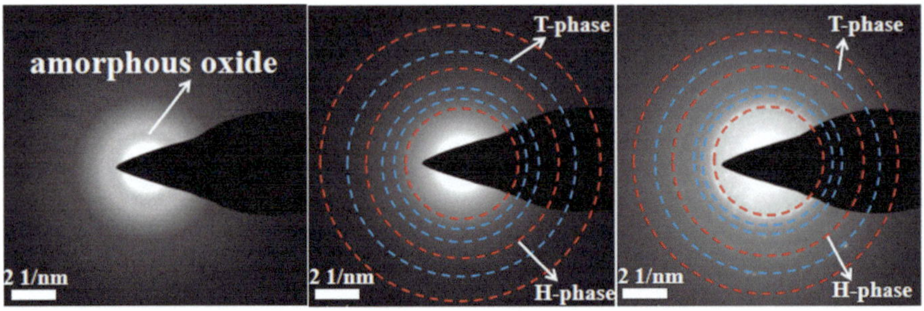

Fig. 8.6 Selected electron diffraction pattern of ZrNbCrAlSiO film/coating with high silicon content after oxidation at different temperatures for 2 h at 1000 °C: **a** after oxidation at 800 °C for 2 h, **b** after oxidation at 900 °C for 2 h, and **c** after oxidation at 1000 °C for 2 h

the average grain size measures approximately 4.19 nm, whereas after oxidation at 1000 °C for 2 h, it increases to around 5.8 nm. This amorphous SiO_2-coated nanocrystalline structure not only effectively suppresses grain growth but also fills crystal defects, obstructs oxygen diffusion along grain boundaries, and exhibits excellent oxygen inhibition properties. Moreover, the formation of amorphous SiO_2 surrounding the corresponding nanocrystals acts as a barrier against the diffusion of other metal atoms and retards oxidation reactions, thereby impeding internal element diffusion. Revesz et al. showed that amorphous oxide coatings are favorable barriers to oxygen ion diffusion due to their dense and boundaryless amorphous structure [20–22]. Cox et al. showed by oxygen isotope diffusion curve analysis that grain-boundary diffusion is the main location where diffusion occurs relative to bulk diffusion [21]. The formation of a dense amorphous protective oxide layer during high-temperature oxidation of coating materials is essential to improve the oxygen-barrier properties of the materials [23–25].

In order to investigate the protective effect of $(ZNCA)_1S_3$ on Zr-4 alloy, the oxidation experiments were conducted at different times in an air environment at 1000 °C, and the thickness of the oxide film was characterized with or without coating protection. Figure 8.7 illustrates the SEM cross-section and in-face images of Zr-4 alloy protected by after oxidation for 0.5 h at 1000 °C in an air environment. Figure 8.7a represents the initial state of the coating, which exhibits a relatively complete and dense structure without any cracks, holes, or other defects. The SEM cross-section topography reveals that the coating is tightly bonded to the substrate of Zr-4 alloy while maintaining its structural integrity without any cracks or holes. The Zr-4 alloy, protected by $(ZNCA)_1S_3$ coating, was subjected to oxidation at 1000°C for 0.5 h in an air environment. No microcracks were observed in the planar graph, indicating that the addition of the ZrSiO protective coating effectively suppressed the formation of surface cracks under high-temperature air conditions. Subsequently, we conducted a detailed analysis of the mechanical properties of the ZrNbCrAlSiO coating to

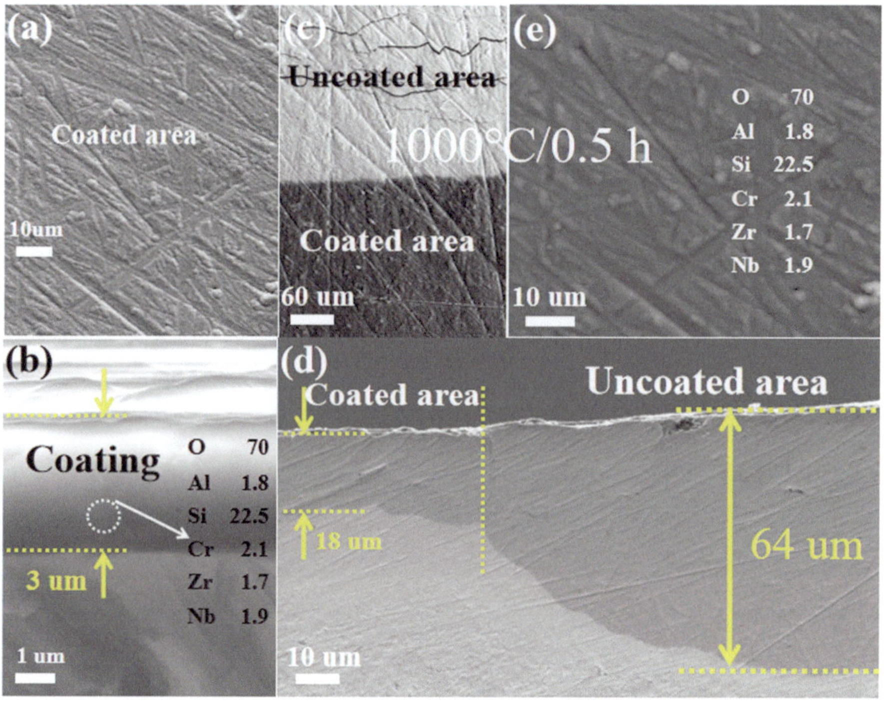

Fig. 8.7 SEM cross-section and plane diagram of prepared Zr-4 alloy protected by $(ZNCA)_1S_3$ and after oxidation at 1000 °C for 0.5 h in air environment: **a** SEM diagram of the prepared coating, **b** SEM cross-section diagram of the prepared coating, **c** SEM diagram at the interface of the coating after oxidation at 1000 °C/0.5 h, **d** SEM cross-section diagram at the interface after oxidation at 1000 °C/0.5 h, **e** SEM diagram of the coating after oxidation at 1000 °C/0.5 h

understand the reasons behind small cracks on its surface. Figure 8.7d shows a SEM cross-section at the interface after oxidation at 1000 °C/0.5 h. It can be observed that without any coating protection, an oxide film with a thickness of approximately 64 μm formed on the Zr-4 alloy. However, $(ZNCA)_1S_3$ protected Zr-4 alloy only exhibited an increase of about 18 μm in oxide film thickness. This result indicates that $(ZNCA)_1S_3$ provides certain protection for the Zr-4 alloy.

Figure 8.8 presents the cross-section and in-plane SEM images of the Zr-4 alloy protected by $(ZNCA)_1S_3$ coating after oxidation for 2 h at 1000°C. Figure 8.8a illustrates that without coating protection, the oxidized surface of the Zr-4 alloy exhibits a crisscrossed structure, while the surface of $(ZNCA)_1S_3$ coating remains relatively intact. Figure 8.8b provides a magnified view of this structure without any observed microcracks present at the coating position. The SEM cross-section topography reveals that without coating protection, a bare zirconium alloy oxide film

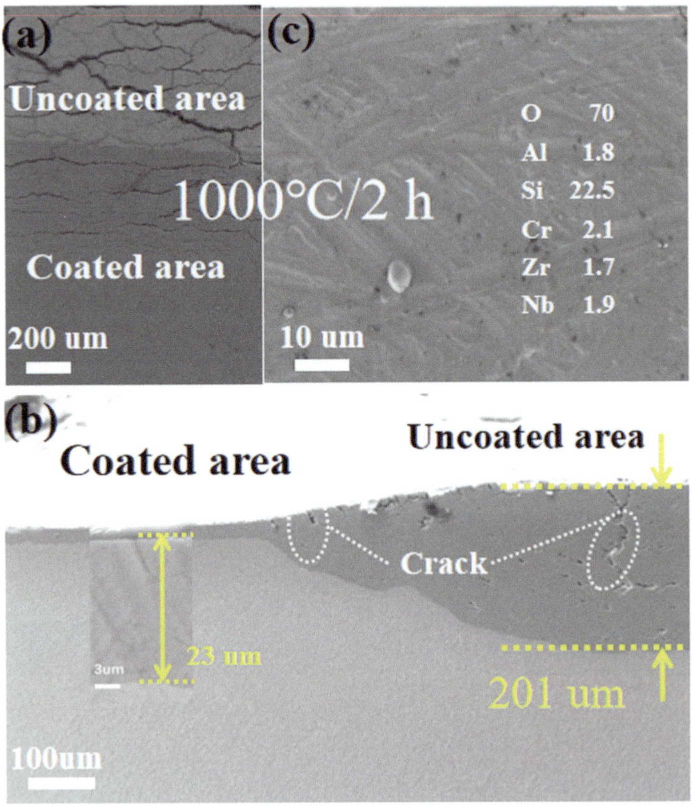

Fig. 8.8 SEM cross-section and plane diagram of Zr-4 alloy protected by $(ZNCA)_1S_3$ after oxidation at 1000 °C for 2 h in air environment: **a** SEM plane diagram at the interface after 1000 °C/2 h of oxidation, **b** SEM cross-section diagram at the interface after 1000 °C/2 h of oxidation, **c** SEM plane diagram of the coating after 1000 °C/2 h of oxidation

forms with a thickness of 201μm after oxidation at 1000 °C for 2 h. However, when protected by $(ZNCA)_1S_3$ coating, only a thin layer approximately measuring about 23μm undergoes oxidation. In Fig. 8.9, the EDS mapping results show that there is no significant internal oxygen diffusion phenomenon, demonstrating good protective effect. In Fig. 8.10, the oxidation kinetics of both the bare Zr-4 alloy and the Zr-4 alloy protected by $(ZNCA)_1S_3$ in a high-temperature air environment at 1000 °C/2 h can be described by an approximate parabolic law. After oxidation at 1000 °C/2 h, the thickness of the oxide film on the bare Zr-4 alloy is measured to be 201 μm, whereas for the Zr-4 alloy protected by a ZrNbCrAlSiO coating, it only increases by 23 μm. Notably, the growth rate of oxide film on Zr-4 alloy with protective coating follows an approximate parabolic law and exhibits significantly lower values.

In order to investigate the cause of microcrack disappearance on the surface of $(ZNCA)_1S_3$, we conducted nanoindentation tests on the coating before and after oxidation at various temperatures. The hardness and elastic modulus of $(ZNCA)_1S_3$ were measured using the continuous stiffness method. Ten indentation points were selected for each coating surface to determine the elastic modulus and hardness. Subsequently, average values along with standard deviations under different temperature oxidation conditions were calculated, enabling us to evaluate the plastic deformation resistance (H^3/E^2) value of $(ZNCA)_1S_3$ based.

The change curve of the elastic modulus of $(ZNCA)_1S_3$ under different oxidation conditions is shown in Fig. 8.11a. It can be observed that the elastic modulus gradually

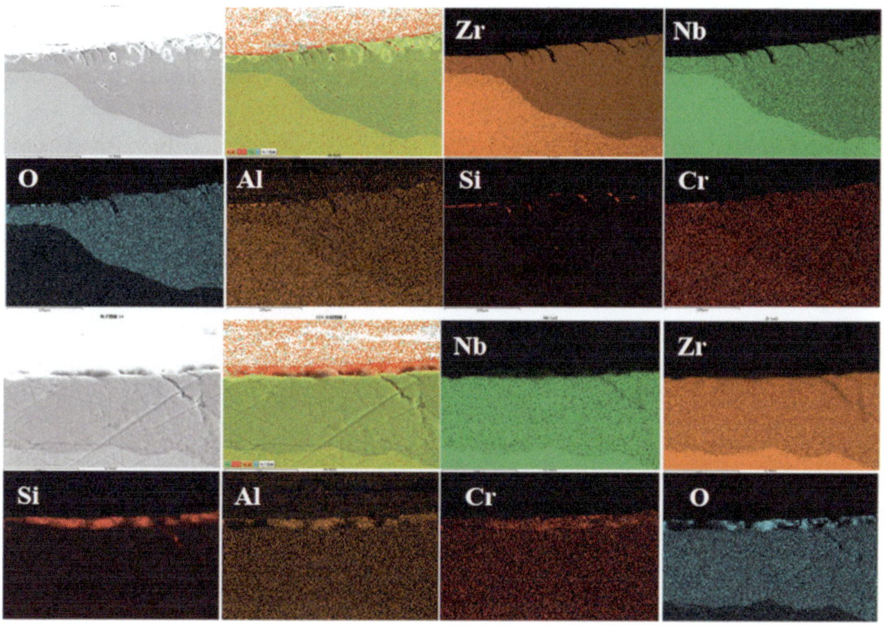

Fig. 8.9 EDS distribution of elements on SEM cross-section of Zr-4 alloy protected by $(ZNCA)_1S_3$ after oxidation at 1000 °C for 2 h in air environment

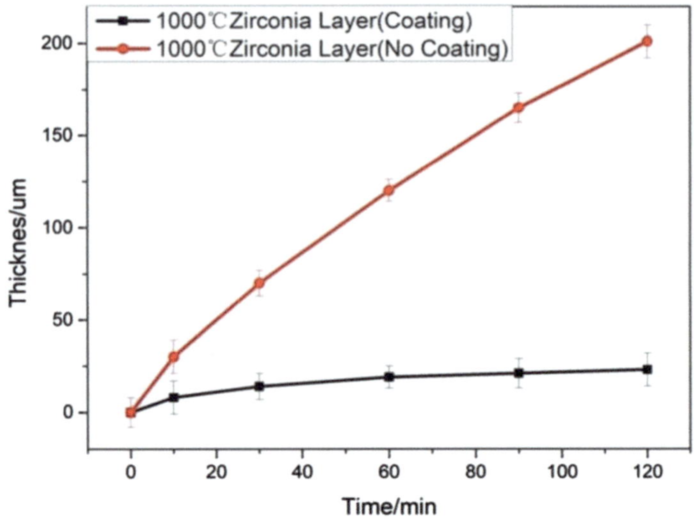

Fig. 8.10 Thickening curve of oxide film of Zr-4 alloy protected by bare Zr-4 alloy and $(ZNCA)_1S_3$-coated Zr-4 alloy at 1000 °C air environment

increases with increasing oxidation temperature. The average elastic modulus of $(ZNCA)_1S_3$ is 92.4 GPa, which rises to 125.6 GPa after oxidation at 1000°C for 2 h. The curve in Fig. 8.11b represents the hardness variation of $(ZNCA)_1S_3$ under different oxidation conditions. It can be observed that the hardness of the coating increases to a certain extent as the oxidation temperature rises. In its initial state, the average hardness of $(ZNCA)_1S_3$ is merely 5.5 GPa, whereas after oxidation at 1000 °C for 2 h, it reaches an average hardness of 9.2 GPa.

The plastic deformation resistance (H^3/E^2) value of $(ZNCA)_1S_3$ under different oxidation conditions is shown in Fig. 8.11d. It can be observed that the plastic deformation resistance (H^3/E^2) value exhibits a gradual increase as the oxidation temperature rises. A higher value of plastic deformation resistance (H^3/E^2) indicates greater resistance to plastic deformation, increased stiffness of the coating, reduced toughness and plasticity, and enhanced susceptibility to surface crack formation and expansion. After being annealed at 1000 °C for 2 h, the Zr–O–Si coating achieves a plastic deformation resistance (H^3/E^2) value of 0.0971 Gpa while the corresponding value for ZrNbCrAlSiO coating is only 0.0494 Gpa—merely half that of the former case after identical oxidation conditions are applied for both coatings at this temperature level. Consequently, it can be concluded that the ZrNbCrAlSiO coating exhibits superior ductility compared to its equivalent silicon content Zr–O–Si counterpart; moreover, when exposed to oxidizing environments at elevated temperatures, it demonstrates reduced propensity for surface crack formation and expansion due to these characteristics. Furthermore, as depicted in Fig. 8.12, upon undergoing high-temperature oxidation treatment, the $(ZNCA)_1S_3$ coating forms a

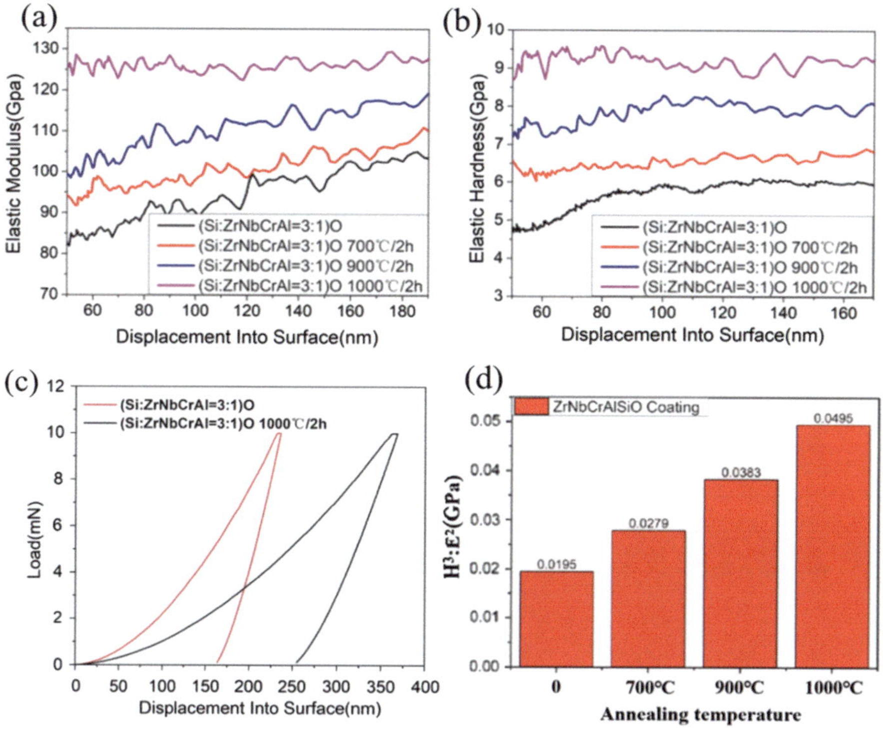

Fig. 8.11 **a** Elastic modulus curve of $(ZNCA)_1S_3$ with pressing depth, **b** the hardness curve of $(ZNCA)_1S_3$ with pressing depth, **c** loading–unloading curve, and **d** value of plastic deformation resistance (H^3/E^2) of the coating after oxidation at different temperatures

structure consisting of amorphous SiO_2 phase coated with high-entropy nanoparticles. Notably within this structure, the presence of these high-entropy nanoparticles imparts improved toughness properties on the overall coating system by effectively hindering crack propagation processes.

The $(ZNCA)_1S_3$ exhibits excellent oxygen diffusion prevention ability and effectively protects the Zr-4 alloy during oxidation in an air environment at 1000 °C. To further investigate the protective performance of the coating, we will increase the test temperature to 1200 °C. Figure 8.12 presents the optical topography of $(ZNCA)_1S_3$-coated Zr-4 alloy after being oxidized for 2 h in an air environment at 1200 °C. At this elevated temperature, the uncoated Zr-4 alloy experienced approximately 320 μm of oxidation, whereas the coated Zr-4 alloy only exhibited minimal oxidation in a thin layer. Figure 8.12c illustrates the interface position topography of both coated and uncoated Zr-4 alloys, clearly demonstrating that the coating plays a significant protective role.

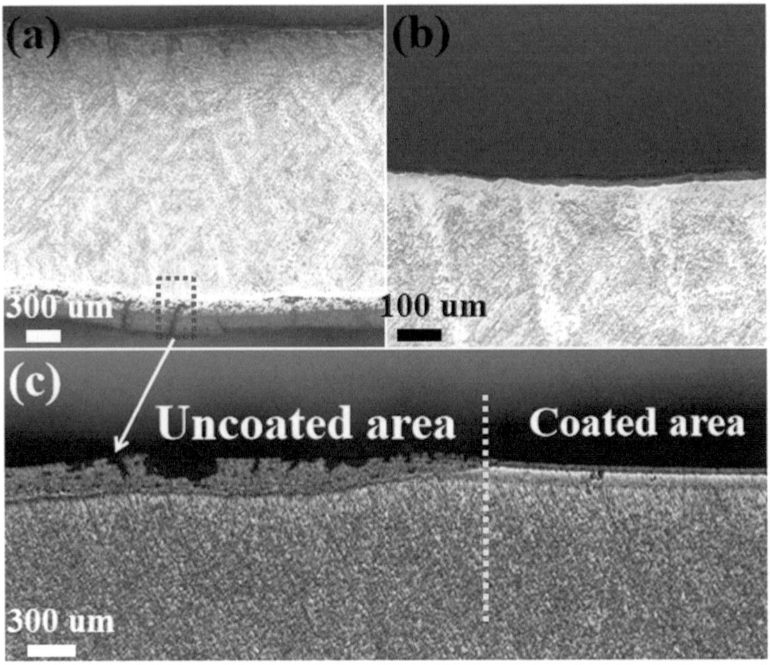

Fig. 8.12 Morphology of Zr-4 alloy protected by $(ZNCA)_1S_3$ after 1200 °C oxidation for 2 h air oxidation

To investigate the protective role of $(ZNCA)_1S_3$ in nuclear reactor water loss accidents, we conducted a simulation experiment at 1200 °C in a water vapor environment. Figure 8.13a displays the morphology of Zr-4 alloy protected by ZrNbCrAlSiO coating after being oxidized for 2 h in a water vapor environment at 1200 °C. The uncoated side's oxide film on the Zr-4 alloy increased to approximately 420 μm, while only a thin layer of oxidation occurred on the coated position. Figure 8.13b is an enlarged view of the coated position on the Zr-4 alloy, showing that only about 35μm of oxide film grew under protection from the coating. Therefore, even in a water vapor environment at 1200 °C, $(ZNCA)_1S_3$ can provide excellent protection.

The variation curve of oxide film thickness for Zr-4 alloy, protected by bare Zr-4 alloy and $(ZNCA)_1S_3$ coating, is shown in Fig. 8.14 as the oxidation time increases in a water vapor environment at 1200 °C. The thickening curve of the oxide film on exposed Zr-4 alloy and the $(ZNCA)_1S_3$ protected by Zr-4 alloy at 1200 °C/2 h in a high-temperature water vapor environment follows an approximate parabolic law. After oxidation at 1200 °C/2 h in a high-temperature steam environment, the oxide film thickness of bare Zr-4 alloy reaches 420 μm. In contrast, the growth of the oxide film on Zr-4 alloy protected by $(ZNCA)_1S_3$ coating is only 35 μm after 2 h of oxidation in a 1200 °C air environment. The thickness curve of the oxide film on Zr-4 alloy protected by $(ZNCA)_1S_3$ coating follows an approximate parabolic rule and is significantly lower than that observed for bare Zr-4 alloy at this stage.

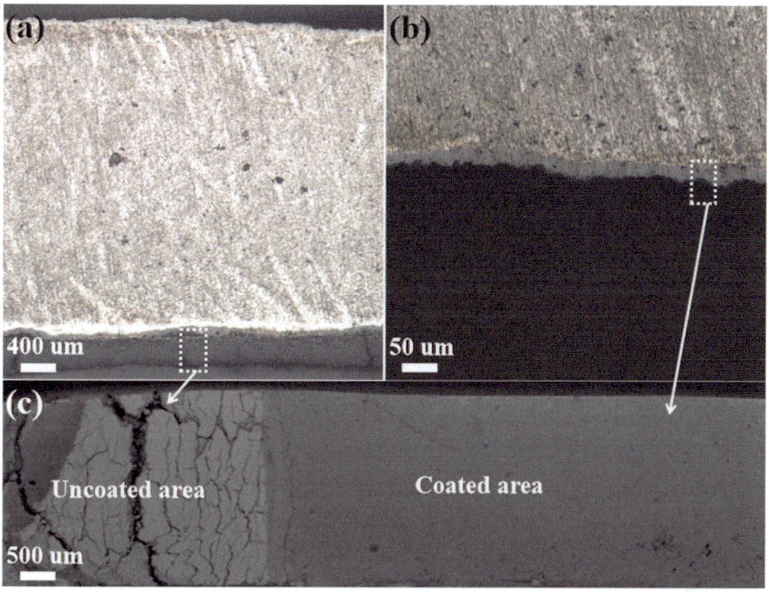

Fig. 8.13 Morphology of Zr-4 alloy protected by $(ZNCA)_1S_3$ coating with or without high silicon content oxidized for 2 h in water vapor environment at 1200 °C. **a** Cross-section morphology of Zr-4 alloy protected by $(ZNCA)_1S_3$ for 2 h in water vapor environment at 1200 °C. **b** Cross-section topography of Zr-4 alloy with coating protection oxidized in water vapor environment at 1200°C for 2 h. **c** Plane topography of the interface position

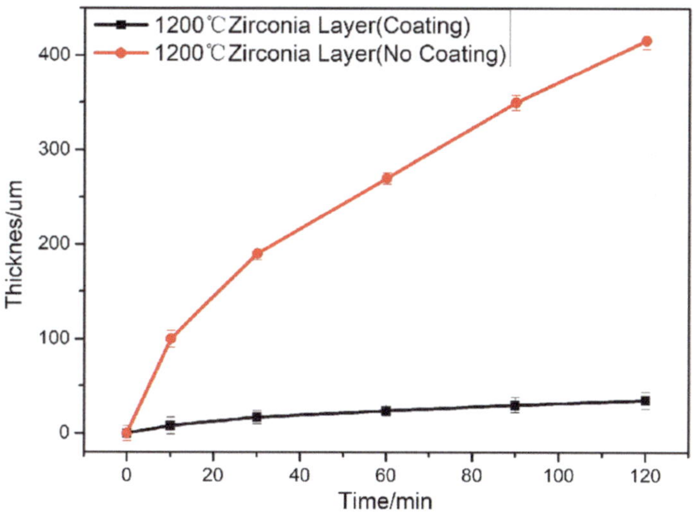

Fig. 8.14 Time thickening curve of the oxide film of Zr-4 alloy protected by bare Zr-4 alloy and $(ZNCA)_1S_3$ coating with high silicon content at 1200 °C water vapor environment

8.4 Study on Water Environment Stability Structure Design and Electrochemical Corrosion of ZrNbCrAlSiO Coating

The susceptibility of silicon-based materials to dissolution in high-temperature and high-pressure water environments is widely acknowledged. In order to further assess the corrosion behavior of $(ZNCA)_1S_3$ coatings in subcritical water environments, we conducted an experiment designed to simulate pressurized water reactor (PWR) conditions. Corrosion experiments were carried out in a 0.01 mol/L LiOH solution at 320 °C and 16 MPa for durations ranging from 5 to 20 h.

In the discussion of 8.1, it is evident that the $(ZNCA)_1S_3$ coating exhibits commendable protective efficacy on Zr-4 alloy in an air environment at 1000 °C. Subsequently, an investigation into the stability of this coating in a subcritical water environment was conducted. Figure 8.15a illustrates the SEM cross-sectional topography of the $(ZNCA)_1S_3$ coating fabricated on Zr-4 alloy, revealing its compact structure and strong adhesion to the substrate.

The SEM cross-section morphology of $(ZNCA)_1S_3$ coating after hydrothermal corrosion in a 0.01 mol/L LiOH solution (320 °C, 16 MPa, 0.01 mol/L) for 5 h is shown in Fig. 8.15c and d. It can be observed from the cross-sectional topography that the coating structure becomes more porous and exhibits partial delamination from the Zr-4 alloy substrate after undergoing hydrothermal corrosion for 5 h. EDS point analysis revealed a decrease in silicon atomic percentage from 22.5% to 14.5%, as well as a decrease in aluminum atomic percentage from 1.8% to 1.0% within the coating

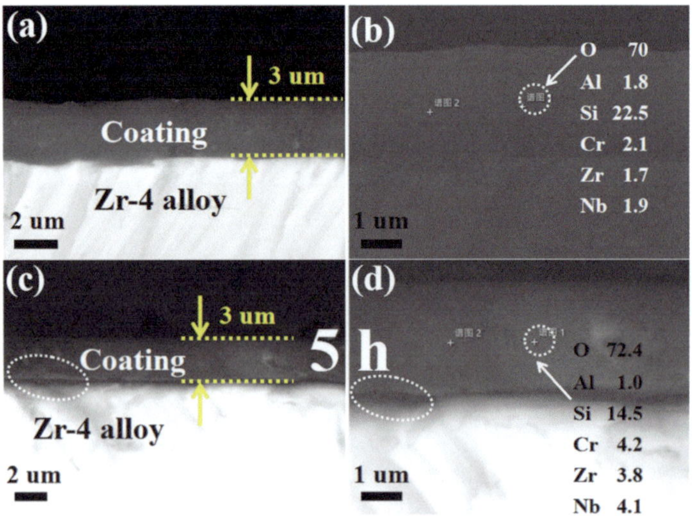

Fig. 8.15 SEM cross-sections of the $(ZNCA)_1S_3$ coating in the prepared state and after 5 h of corrosion in LiOH solution at high temperature and pressure (320 °C, 16 MPa, 0.01 mol/L): **a**, **b** cross-sections in the prepared state **c**, **d** cross-sections after 5 h of corrosion

following a duration of hydrothermal corrosion for five hours, indicating reduced stability of these elements within the coating. The silicon and aluminum elements in the coating are dissolved in water under subcritical conditions, respectively, by the reactions (8.1) and (8.2) [26].

$$SiO_2 + 2H_2O = SiO(OH)_3^- + H^+, \tag{8.1}$$

$$Al + 2H_2O \rightarrow AlO(OH) + 3/2H_2. \tag{8.2}$$

Based on the aforementioned discussion, it is evident that the $(ZNCA)_1S_3$ coating cannot maintain stability and provide effective protection in a 0.01mol/L LiOH solution at 320 °C and 16 Mpa. To address this issue and develop a coating structure capable of safeguarding Zr-4 alloy not only in an air environment at 1000 °C but also in a subcritical water environment while maintaining structural integrity and compositional stability, the surface layer of the $(ZNCA)_1S_3$ coating was subjected to low-oxygen modification. The $(ZNCA)_1S_3$ coating was subjected to surface low-oxygen modification, resulting in a suboxidized surface layer of the $(ZNCA)_1S_3$ coating, while the normally prepared $(ZNCA)_1S_3$ coating closely resembled the Zr-4 alloy base. Figure 8.16 presents the DES point measurement results of the surface low-oxygen-modified $(ZNCA)_1S_3$ coating, revealing that only 26% atomic percentage of surface oxygen is present (compared to over 65% for completely oxidized oxygen), indicating an oxygen-deficient state. Additionally, DES surface scan element distribution confirms a low level of oxygen content in the surface coating. Figure 8.17 displays SEM cross-section morphology and element distribution of the surface low-oxygen-modified $(ZNCA)_1S_3$ coating after undergoing corrosion for 20 h in a high-temperature and high-pressure LiOH solution (320 °C, 16 MPa, 0.01 mol/L). The DES point scanning reveals that the atomic percentage of surface coating oxygen reaches 71%, transforming it into a fully oxidized state. Meanwhile, there is no decrease observed in silicon and aluminum elements' relative content within the coating; these elements remain stable within this structure. Furthermore, all elements are evenly distributed according to DES surface scan analysis, with no oxidation detected on Zr-4 alloy.

The surface optical morphology of $(ZNCA)_1S_3$ coating and surface hypoxy-modified $(ZNCA)_1S_3$ coating before and after hydrothermal corrosion in LiOH solution (320 °C, 16 MPa, 0.01 mol/L) at high temperature and pressure is shown in Fig. 8.18. After undergoing high-temperature and high-pressure hydrothermal corrosion for 5 h, the local surface of the $(ZNCA)_1S_3$ coating exhibits reduced density; furthermore, after a duration of 20 h under hydrothermal corrosion conditions, the entire coated surface becomes loose and pulverized. In contrast, the low-oxygen-modified $(ZNCA)_1S_3$ coating maintains its dense surface structure even after being subjected to high-temperature and high-pressure hydrothermal corrosion for 20 h, indicating its stable existence and protective role in a LiOH hydrothermal environment under such extreme conditions.

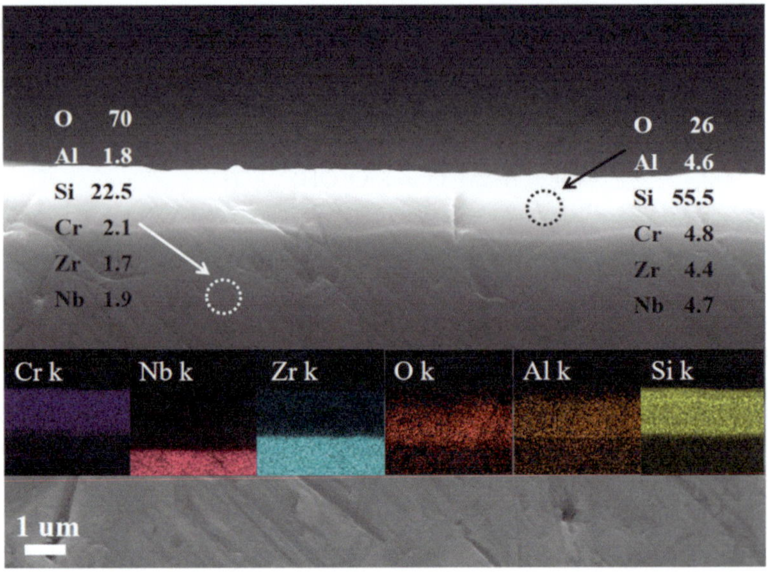

Fig. 8.16 SEM cross-sectional morphology and elemental distribution of the prepared state of the surface low-oxygen $(ZNCA)_1S_3$ coating

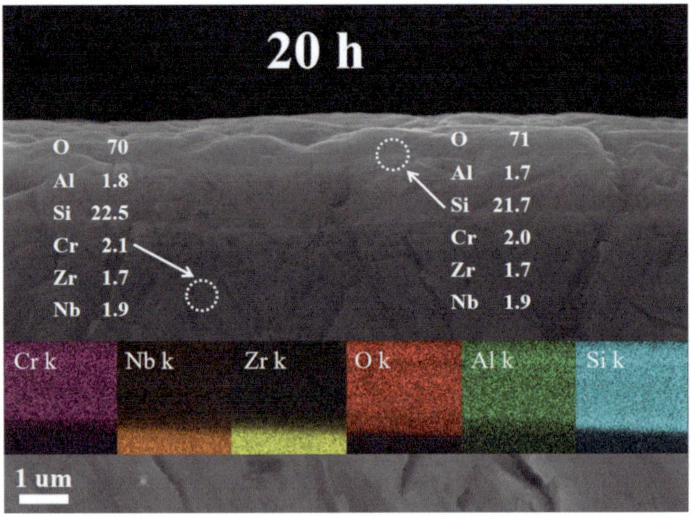

Fig. 8.17 SEM cross-sectional morphology and elemental distribution of the surface low-oxygen-modified $(ZNCA)_1S_3$ coating after 20 h of corrosion in high-temperature and high-pressure LiOH solution (320 °C, 16 MPa, 0.01 mol/L)

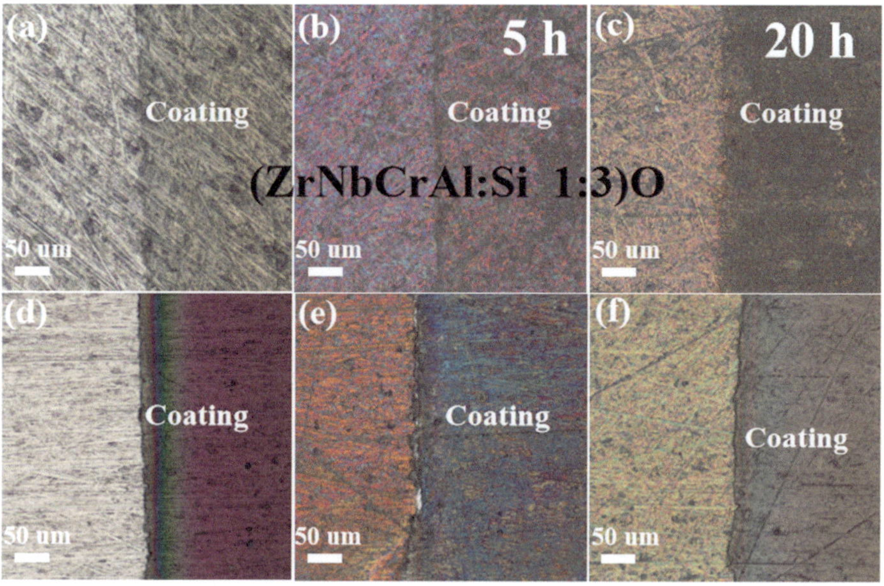

Fig. 8.18 Surface optical morphology of $(ZNCA)_1S_3$ coating and superficial low-oxygen-modified $(ZNCA)_1S_3$ coating before and after hydrothermal corrosion in high-temperature and high-pressure LiOH solution (320 °C, 16 MPa, 0.01 mol/L): **a–c** surface morphology of high silicon content coating preparation, hydrothermal 5 h, hydrothermal 20 h respectively **d–f** surface morphology of superficial low-oxygen coating preparation, hydrothermal 5 h, hydrothermal 20 h respectively **f** Surface morphology of the surface layer with low-oxygen content in the prepared state, hydrothermal 5 h and hydrothermal 20 h, respectively

The XPS fitting results of the $(ZNCA)_1S_3$ coating in its prepared state and after corrosion in a LiOH solution (320 °C, 16 MPa, 0.01 mol/L) at high temperature and pressure for 5 h are shown in Fig. 8.19 (the top curve in each Figure represents the XPS peak fitting result of the element's prepared state, while the bottom curve represents the XPS peak fitting result of the element's hydrothermal corrosion for 5 h). The prepared state of the $(ZNCA)_1S_3$ coating is fully oxidized, with a fitted Si peak at 102.8 eV corresponding to Si 2p characteristic peak of SiO_x, which closely matches the binding energy of SiO_2 [27, 28]. The fitted Al peak at 74.4 eV corresponds to Al 2p characteristic peak of Al_2O_3 [29] observed in its prepared state. Similarly, Zr exhibits a main fitted peak at 183.0 eV corresponding to Zr 3d5/2 characteristic peak of ZrO_2 [30], while Nb shows a main fitted peak at 207.6 eV corresponding to Nb 3d5/2 characteristic peak of Nb_2O_3 [31]. The main fitting peak of the prepared Cr is observed at a position of 577.3 eV, corresponding to the characteristic Cr 2p3/2 peak of Cr_2O_3 [32]. Following hydrothermal corrosion, a significant loss of silicon and aluminum elements occurs, resulting in the presence of two predominant bonding states for remaining Si: Si–OH and Si–O bonds; similarly, two dominant bonding states are observed for remaining Al: Al–OH and Al–O bonds. Additionally, metal elements after hydrothermal corrosion

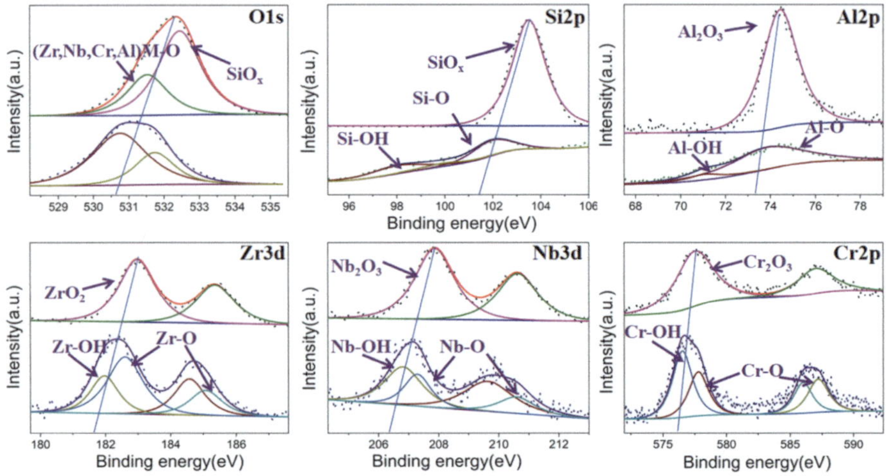

Fig. 8.19 XPS fitting results for the $(ZNCA)_1S_3$ coating in the prepared state and for each element after 5 h of corrosion in high-temperature and high-pressure LiOH solution (320 °C, 16 MPa, 0.01 mol/L)

exist in two bonding states: M(Zr, Nb, Cr, Al)-OH and M(Zr, Nb, Cr, Al)-O. The binding energy values for silicon and other metal elements decrease compared to their prepared state counterparts following hydrothermal corrosion. Due to the substantial reduction in silicon and aluminum contents after sample's hydrothermal corrosion process, oxygen exhibits lower valence states within the multicomponent structure post-corrosion with correspondingly smaller binding energy values.

The XPS fitting results of the prepared surface hypoxy-modified $(ZNCA)_1S_3$ coating and each element after corrosion in a LiOH solution (320 °C, 16 MPa, 0.01 mol/L) at high temperature and pressure for 20 h are shown in Fig. 8.20 (the top curve in each Figure represents the XPS peak fitting result of the initial state of the element, while the bottom curve represents the XPS peak fitting result of hydrothermal corrosion of the element for 20 h). The $(ZNCA)_1S_3$ coating in the suboxide state exhibits a surface with low-oxygen content. Owing to the high electronegativity of Si and Al, the metal elements (Zr, Nb, Cr) form bonds not only with oxygen but also with silicon and aluminum. As there is no evident Si–O or Al-O bond present in the prepared coating state, it remains relatively stable under conditions of high-temperature and pressure water without undergoing any dissolution reaction. In this oxidizing state, the coating combines with oxygen released from water at elevated temperatures and pressures to eventually transform into a fully oxidized state. When oxygen enters the coating, it tends to combine with metal elements (Zr, Nb, Cr, Al). Upon combining with these metal elements, oxygen forms a mixed bonding state of M(Zr, Nb, Cr, Al)-O-Si-, which characterizes the formation of this mixed bonding state. Consequently, the binding energy of Si (102.9 eV) slightly increases but remains smaller than that in silicon oxide (103.75 eV). Simultaneously, the binding energy (103.5 eV) of silicon in ZrNbCrAlSiO coating with high silicon content is also lower

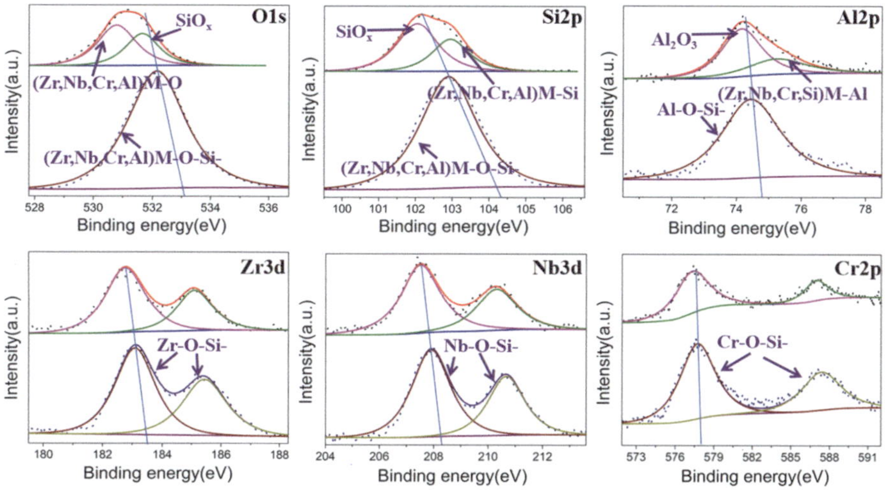

Fig. 8.20 XPS fitting results of the surface low-oxygen-modified $(ZNCA)_1S_3$ coating in the prepared state and after 20 h of corrosion in high-temperature and high-pressure LiOH solution (320 °C, 16 MPa, 0.01 mol/L) for each element

compared to that in ZRNbCralsio coating. This complex bonding state exhibits a denser atomic structure and effectively impedes the ingress of oxygen and water molecules. Following high-temperature and high-pressure hydrothermal corrosion treatment, more oxygen combines with metal elements within the coating to form a mixed bonding state of M(Zr, Nb, Cr, AI)-O-Si-. As a result, the binding energy for each metal element increases to some extent relative to its initial prepared state. Furthermore, this non-simple silica tetrahedral structure enhances resistance against water molecule penetration.

The corrosion resistance of ZrNbCrAlSiO coatings with different compositions was investigated under electrochemical conditions. Figure 8.21 shows the kinetic potential polarization curves of bare Zr-4 alloy and ZrNbCrAlSiO coatings with different silicon contents in 0.01 mol/L LiOH solution. The ZrNbCrAlSiO coatings with three different silicon metal atom ratios also have better corrosion resistance than the bare Zr-4 alloy. The corrosion potential of the bare Zr-4 alloy is −0.865 V, the corrosion potentials of the samples with ZrNbCrAlSiO coatings are shifted to the right to varying extents, and the corrosion potential of the $(ZNCA)_1S_3$ coating is close to −0.277 V. The self-corrosion current density of the bare Zr-4 alloy is 0.01 mol/L LiOH solution. 4 alloy was 6.766×10^{-8} A cm^{-2}, and the self-corrosion current density of the coated samples was reduced to different degrees compared with that of the bare Zr-4 alloy, and the self-corrosion current density of the $(ZNCA)_1S_3$-coated samples was only 5.316×10^{-10} A cm^{-2}, which indicated that the ZrNbCrAlSiO coating had a better corrosion resistance than that of the bare Zr-4 alloy. ZrNbCrAlSiO coating has better corrosion resistance than bare Zr-4 alloy. In the early stage of anodic polarization, the current density of bare Zr-4 alloy increases rapidly with the increase

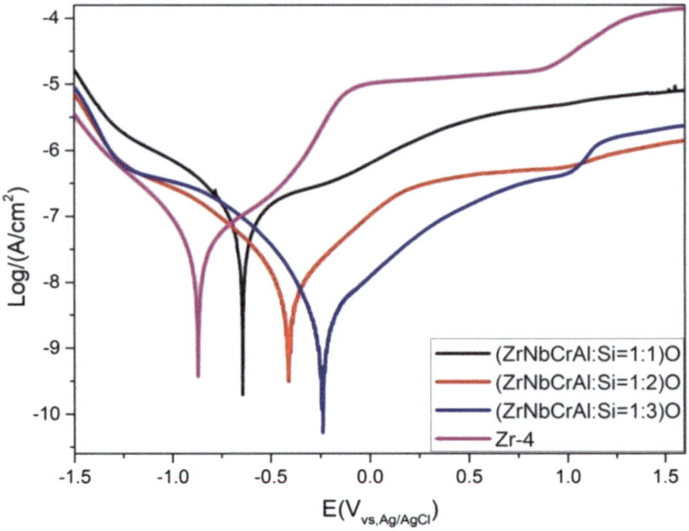

Fig. 8.21 Polarization curves of Zr-4 alloy substrates and ZrNbCrAlSiO coatings of different compositions in 0.01 mol/L LiOH solution

of potential, while the current density of ZrNbCrAlSiO-coated samples increases slowly with the increase of potential in the early stage of anodic polarization. This is due to the bare Zr-4 alloy corrosion occurs preferentially at the location of grain boundaries and crystal defects, while for the amorphous ZrNbCrAlSiO coating due to the absence of grain boundaries, dislocations, second particles, and other defects, corrosive electrolytes are difficult to penetrate into the coating, increasing the corrosion potential and delaying the corrosion rate, and effectively improve the corrosion resistance of the coating. In addition, ZrNbCrAlSiO coating consists of corrosion-resistant elements (Zr, Cr, Al, Si), which can quickly form a passivation film to further enhance the blocking effect on the electrolyte solution and improve the corrosion resistance of the coating.

The optical microscopic morphology of Zr-4 alloy and ZrNbCrAlSiO coatings of different compositions after electrochemical corrosion is shown in Fig. 8.22. The bare Zr-4 alloy showed many pitting pits after electrochemical corrosion, while the three-composition ZrNbCrAlSiO-coated samples showed only very few corrosion defects. Compared with the transmitted Zr-4 alloy, the amorphous ZrNbCrAlSiO coating has a unique long and short disordered atomic arrangement without corrosion-sensitive defects such as grain boundaries, dislocations, and second phases, and its corrosion resistance is higher than that of the Zr-4 alloy.

Figure 8.23 shows the electrochemical impedance curves of ZrNbCrAlSiO coatings with different silicon contents in 0.01 mol/L LiOH solution. In the impedance plot, the size of the radius of the fitted impedance ring correlates with the corrosion resistance of the test sample. After obtaining the raw data, Zview2 software was used to establish an equivalent circuit model for fitting, and the error was controlled

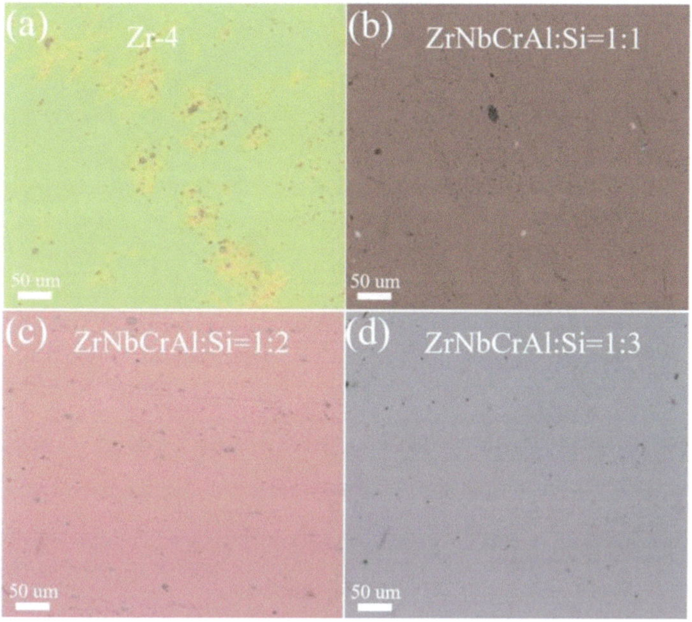

Fig. 8.22 Optical microstructure of samples after electrochemical corrosion of Zr-4 and ZrNbCrAlSiO coatings with different silicon contents

within 2% during the fitting process to ensure that the fitting results were reliable. The fitting parameters are shown in Table 8.1, where Rs is the solution resistance, the value of which depends on the distance from the working electrode (test sample) to the reference electrode in the lithium hydroxide solution, which is kept constant during the test, so Rs is a stable value. Rc is the physical barrier property of the coating, i.e., the coating resistance. CPE is the constant phase element. The electrochemical capacitance of CPE is calculated using Eq. (8.3):

$$C_{\mathrm{CPE}} = \frac{(QR_c)^{1/n}}{R_c}, \tag{8.3}$$

where R is the corresponding resistance. The calculated CCPE values are presented in Table 8.1.

Table 8.1 shows the results of fitting the electrochemical impedance, and the results show that the n values of all the samples are very similar and close to 1, indicating that all the coatings and Zr-4 alloys have almost the same surface flatness. A higher CCPE value usually accelerates the rate of charge exchange on the electrode surface to accelerate the dissolution process of the test samples, at this time, the surface structure is destroyed, and the corrosion resistance will be reduced; the three different compositions of the ZrNbCrAlSiO coatings of three different compositions all have smaller CCPE values than Zr-4 alloy, so the coated samples all have stronger

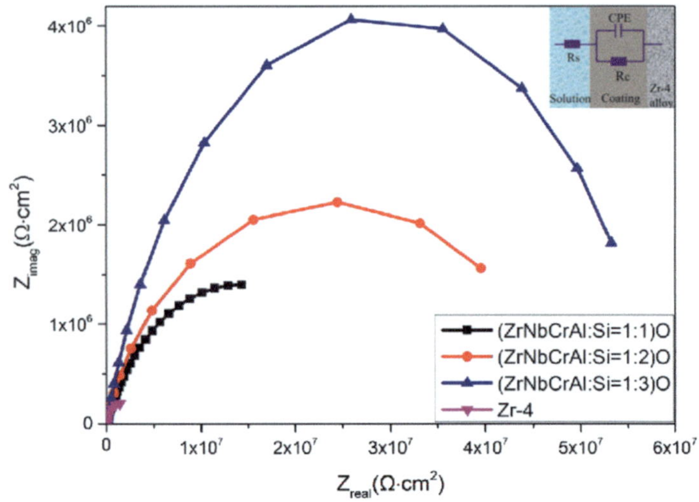

Fig. 8.23 EIS Nyquist curves for different composition ZrNbCrAlSiO-coated samples in 0.01 mol/L LiOH solution, inset shows the equivalent circuit physical model of the coating, R_s = solution resistance, R_c = coating resistance, CPE = coating capacitance

Table 8.1 Main parameters of the electrochemical impedance test

Samples	R_s (Ωcm^2)	$Q(\Omega^{-1}\ s^n\ cm^{-2})$	n	R_c (Ωcm^2)	C_{CPE} (F cm^{-2})
Zr-4	8.24	2.44×10^{-6}	0.91	3.19×10^5	2.38×10^{-6}
$(ZNCA)_1S_1$	9.82	1.78×10^{-7}	0.89	2.82×10^7	3.05×10^{-7}
$(ZNCA)_1S_2$	11.2	7.91×10^{-8}	0.88	4.79×10^7	1.02×10^{-7}
$(ZNCA)_1S_3$	8.6	2.87×10^{-8}	0.92	5.84×10^7	3.32×10^{-8}

corrosion resistance than Zr-4 alloy. The coating materials all have larger Rc values than Zr-4 alloy, and the Rc values tend to increase with the increase of silicon content, which indicates that the ZrNbCrAlSiO coatings have higher physical barrier capacity with the increase of silicon content.

8.5 Conclusion

ZrNbCrAlSiO coatings with different atomic ratios were deposited on Zr-4 alloy to investigate the protective effect of ZrNbCrAlSiO coatings on Zr-4 alloy at different temperatures. The structural stability as well as the corrosion behavior of the ZrNbCrAlSiO coatings was investigated by high-temperature and high-pressure hydrothermal corrosion experiments and electrochemical corrosion experiments.

(1) ZrNbCrAlSiO coatings with different compositions were prepared and subjected to high-temperature experiments. The amorphous nanocrystalline structure formed after high temperature of $(ZNCA)_1S_3$ coatings can effectively inhibit the growth of nanocrystalline grains and show a better effect of oxygen blocking, and the structure of nanocrystalline grains can alleviate the stresses generated at high temperatures and thus to a certain extent can impede the formation of surface cracks in the case of high-temperature oxidation.

(2) $(ZNCA)_1S_3$ Coating on the Protection of Zr-4 Alloy: The oxide film thickening curves for the Zr-4 alloy with and without coating protection follow parabolic kinetics during oxidation at 1000 °C in air for 2 hours. Compared with the uncoated sample, the coated sample shows a significantly lower oxide film growth rate, demonstrating the coating's excellent protective effect on the Zr-4 alloy. Under high-temperature oxidation conditions, the (ZNCA)1S3 coating exhibits an increasing trend in plastic deformation resistance (quantified by H^3/E^2). However, its H^3/E^2 values remain significantly lower than those of ZS13 coatings subjected to the same treatment. Lower H^3/E^2 values indicate reduced resistance to plastic deformation, corresponding to higher coating plasticity. This enhanced plasticity helps prevent crack formation on the coating surface during service; plastic deformation resistance (H^3/E^2) value is smaller, the smaller the resistance to plastic deformation resistance, the higher the plasticity of the coating, and the coating surface is not easy to form cracks.

(3) Study of Corrosion Resistance of Multicomponent Oxide Coatings: Experimental results show that (ZNCA)1S3 coatings exhibit poor stability in high-temperature, high-pressure LiOH solutions.After hydrothermal corrosion, After hydrothermal corrosion, the surface low-oxygen-modified (ZNCA)1S3 coating forms a (Zr, Nb, Cr, Al)M-O-Si mixed bonding state. This chemically stable structure enables the coating to maintain integrity and protect the underlying Zr-4 alloy under harsh high-temperature (320°C) and high-pressure (16 MPa) LiOH conditions. The polarization and impedance diagrams of the coatings with different Si contents show that the corrosion resistance of the coatings increases with the increase of Si content.

References

1. King DJM, Knowles AJ, Bowden D et al (2022) High temperature zirconium alloys for fusion energy. J Nucl Mater 559:153431
2. Koo YH, Yang JH, Park JY et al (2014) KAERI's development of LWR accident-tolerant fuel. Nucl Technol 186(2):295–304
3. Cao G, Yang L, Yuan G et al (2020) Chemical diversity of iron species and structure evolution during the oxidation of C14 Laves phase Zr $(Fe, Nb)_2$ in subcritical environment. Corros Sci 162:108218
4. Park DJ, Kim HG, Jung YI et al (2016) Behavior of an improved Zr fuel cladding with oxidation resistant coating under loss-of-coolant accident conditions. J Nucl Mater 482:75–82
5. Zinkle SJ, Terrani KA, Gehin JC et al (2014) Accident tolerant fuels for LWRs: A perspective. J Nucl Mater 448(1–3):374–379

6. Kim HG, Kim IH, Jung YI et al (2016) Development of surface modified Zr cladding by coating technology for ATF. Top fuel 2016:1157–1163
7. Tang C, Stueber M, Seifert HJ et al (2017) Protective coatings on zirconium-based alloys as accident-tolerant fuel (ATF) claddings. Corros Rev 35
8. Sidelev DV, Kashkarov EB, Grudinin VA et al (2022) High-temperature oxidation of Cr-coated laser beam welds made from E110 zirconium alloy. Corros Sci 195:110018
9. Dabney T, Johnson G, Yeom H et al (2019) Experimental evaluation of cold spray FeCrAl alloys coated zirconium-alloy for potential accident tolerant fuel cladding. Nuclear Mater Energy 21:100715
10. Wu K, Yao H, Cheng X et al (2020) Oxidation behavior and chemical evolution of architecturally arranged Zr/Si multilayer at high temperature. Surf Coat Technol 399:126205
11. Fabris S, Paxton AT, Finnis MW (2002) A stabilization mechanism of zirconia based on oxygen vacancies only. Acta Mater 50(20):5171–5178
12. Raza M, Cornil D, Cornil J et al (2016) Oxygen vacancy stabilized zirconia (OVSZ); ajoint experimental and theoretical study. Scripta Mater 124:26–29
13. Ramo DM, Shluger AL (2008) Structure and spectroscopic properties of oxygen divacancy in yttrium-stabilized zirconia. J Phys Conf Ser 117(1):012022
14. Cong Y, Li B, Yue S et al (2009) Effect of oxygen vacancy on phase transition and photoluminescence properties of nanocrystalline zirconia synthesized by the onepot reaction. J Phys Chem C 113(31):13974–13978
15. Musil J, Šatava V, Zeman P et al (2009) Protective Zr-containing SiO_2 coatings resistant to thermal cycling in air up to 1400 °C. Surf Coat Technol 203(10–11):1502–1507
16. Roussel M, Talbot E, Pareige P et al (2013) Influence of the supersaturation on Si diffusion and growth of Si nanoparticles in silicon-rich silica. J Appl Phys 113(6):063519
17. Jäger N, Meindlhumer M, Zitek M et al (2022) Impact of Si on the high-temperature oxidation of AlCr (Si) N coatings. J Mater Sci Technol 100:91–100
18. Al-Olayyan Y, Fuchs GE, Baney R et al (2005) The effect of Zircaloy-4 substrate surface condition on the adhesion strength and corrosion of SiC coatings. J Nucl Mater 346(2–3):109–119
19. Usui T, Sawada A, Amaya M et al (2015) SiC coating as hydrogen permeation reduction and oxidation resistance for nuclear fuel cladding. J Nucl Sci Technol 52(10):1318–1322
20. Revesz AG, Fehlner FP (1981) The role of noncrystalline films in the oxidation and corrosion of metals. Oxid Met 15(3–4):297–321
21. Cox B, Pemsler JP (1968) Diffusion of oxygen in growing zirconia films. J Nucl Mater 28(1):73–78
22. Lim KR, Park JM, Kim SJ et al (2013) Enhancement of oxidation resistance of the supercooled liquid in Cu-Zr-based metallic glass by forming an amorphous oxide layer with high thermal stability. Corros Sci 66:1–4
23. Nakamura R, Da To T, Tsukui S et al (2014) Diffusion of oxygen in amorphous Al_2O_3, Ta_2O_5, and Nb_2O_5. J Appl Phys 116(3):222904
24. Weller K, Wang Z, Jeurgens LPH et al (2015) Thermodynamics controls amorphous oxide formation: exclusive formation of a stoichiometric amorphous $(Al_{0.33}Zr_{0.67})O_{1.83}$ phase upon thermal oxidation of Al-Zr. Acta Mater 94:134–142
25. Reichel F, Jeurgens LPH, Mittemeijer EJ (2008) The thermodynamic stability of amorphous oxide overgrowths on metals. Acta Mater 56(3):659–674
26. Kim WJ, Hwang HS, Park JY (2002) Corrosion behavior of reaction-bonded silicon carbide ceramics in high-temperature water. J Mater Sci Lett 21(9):733–735
27. Sawyer R, Nesbitt HW, Secco RA (2012) High resolution X-ray photoelectron spectroscopy (XPS) study of K_2O–SiO_2 glasses: Evidence for three types of O and at least two types of Si. J Non-Cryst Solids 358(2):290–302
28. Oliveira C, Galindo RE, Palacio C et al (2011) Surface characterization of Ti-Si-C-ON coatings for orthopedic devices: XPS and Raman spectroscopy. Solid State Sci 13(1):95–100
29. Zhang Y, Zhang K, Lei S et al (2022) Formation and oxidation behavior of TiO_2 modified Al_2O_3-Nb_2O_5/$NbAl_3$ composite coating prepared by two-step methods. Surface and Coatings Technol 128081

30. Gondal MA, Fasasi TA, Baig U et al (2018) Effects of oxidizing media on the composition, morphology and optical properties of colloidal zirconium oxide nanoparticles synthesized via pulsed laser ablation in liquid technique. J Nanosci Nanotechnol 18(6):4030–4039
31. Malinovschi V, Marin A, Negrea D et al (2018) Characterization of Al_2O_3/ZrO_2 composite coatings deposited on Zr-2.5 Nb alloy by plasma electrolytic oxidation. Appl Surface Sci 451:169–179
32. Zhao Y, Liu W, Fan Y et al (2020) Effect of Cr content on the passivation behavior of Cr alloy steel in a CO_2 aqueous environment containing silty sand. Corros Sci 168:108591

Summary and Conclusions

Chapter 1

This chapter provides an overview of the development of nuclear energy, its role in addressing global energy challenges, and the importance of fuel cladding materials in ensuring the safety and efficiency of nuclear reactors. As the demand for clean and sustainable energy grows, nuclear power, with its low environmental impact and high energy density, has become a critical component of the global energy mix. The evolution of nuclear reactor technology has progressed through four generations, from early military applications to advanced Generation IV reactors, which focus on improving sustainability, safety, and reducing nuclear proliferation risks.

A key safety concern in nuclear reactor design is the performance of fuel cladding materials, which act as the first line of defense against the release of radioactive substances. Zirconium alloys, particularly Zr-4, have long been the material of choice due to their favorable neutron absorption properties and corrosion resistance. However, these materials face significant challenges in terms of oxidation resistance, hydrogen absorption, and mechanical stability under high-temperature conditions, which could undermine reactor safety, particularly during extended operation or in the event of accidents.

To address these challenges, the development of accident-tolerant fuel (ATF) cladding materials has become a priority. ATFs are designed to improve the oxidation resistance and mechanical properties of cladding materials at elevated temperatures, reducing hydrogen generation and heat release during severe accidents such as Loss-of-Coolant Accidents (LOCA). These advancements aim to mitigate the risks of core meltdown and provide additional time for emergency response, enhancing the overall safety of nuclear power plants.

In conclusion, while nuclear energy remains a key solution to global energy demands, ensuring its safety under extreme conditions is crucial for its long-term viability. This chapter underscores the importance of ongoing research into advanced

J. Hu and G. Cao, *Zirconium Alloy Coatings*, https://doi.org/10.1007/978-981-96-5028-6

cladding materials, particularly accident-tolerant fuels, to enhance the safety and reliability of nuclear reactors and support the continued development of nuclear power as a sustainable and secure energy source.

Chapter 2

This chapter focuses on the development of protective coatings for zirconium (Zr) alloy fuel cladding, which are essential for enhancing the safety and performance of nuclear reactors, especially under extreme conditions such as Loss-of-Coolant Accidents (LOCA). The coatings aim to improve oxidation resistance in accident-tolerant fuel (ATF) technology. The key strategies explored include the introduction of novel antioxidant materials and surface modifications of existing Zr alloys:

Chromium (Cr) Coatings: These coatings form a dense Cr_2O_3 layer, offering excellent oxidation resistance. However, prolonged exposure causes degradation of the layer, and the formation of brittle Cr–Zr phases, along with the volatility of CrO_3 at high temperatures, is challenges. Alloying with Al can enhance long-term stability.

FeCrAl Coatings: These alloys form a protective Al_2O_3 layer but suffer from interdiffusion with Zr, leading to brittle Fe–Zr compounds that compromise cladding integrity. A ZrO_2 buffer layer can mitigate this issue.

High-Entropy Alloy (HEA) Coatings: HEAs exhibit high hardness, wear resistance, and thermal stability, with promising oxidation resistance. However, challenges such as complex preparation, high costs, and instability of oxide layers hinder their broader application.

Ceramic Coatings: Composed of silicides, nitrides, and oxides, coatings like CrSi and CrAlN provide good oxidation resistance. However, their performance is limited in high-pressure water environments, and issues like brittleness and nitrogen volatilization at elevated temperatures lead to failure.

MAX-Phase Coatings: Coatings such as Ti_2AlC and Cr_2AlC offer a combination of metallic conductivity and ceramic stability, providing good oxidation protection. However, the depletion of aluminum and formation of voids at high temperatures limit their effectiveness.

In conclusion, while Cr, FeCrAl, HEA, ceramic, and MAX-phase coatings show significant potential, challenges related to material stability, high-temperature diffusion, and cost-effectiveness need to be addressed. Further research is required to optimize these coatings for practical application in advanced nuclear reactors.

Chapter 3

This chapter discusses a variety of surface coating technologies and their applications in the preparation of zirconium alloy coatings, including spraying, arc ion plating, laser cladding, and magnetron sputtering. The coating deposition rate of spraying is fast and it is easy to prepare thick film, but the disadvantage is high porosity and loose structure; the film adhesion of arc ion plating is high and the film is dense and low porosity, but there are the problems of large particles deposition and non-uniformity of sputtering of coating; the fast cooling and heating of laser cladding technology in the process of preparation will form cracks in the coating, which restricts the application of its zirconium alloy coatings; compared to the other coating preparation technologies, magnetron sputtering technology is more suitable for zirconium alloy coatings because of its targeting technology, which is more suitable for the application of laser cladding. Compared with other coating preparation techniques, magnetron sputtering technology has been widely used in the preparation of zirconium alloy cladding coatings due to its wide range of target material selection, small workpiece deformation, controllable deposition rate, good process repeatability, and many other advantages and has been widely used in industrial applications.

Chapter 4

This chapter explores the oxidation behavior and self-strengthening mechanism of Zr–Si amorphous coatings in subcritical water environments. The application of amorphous metallic Si in oxidizing environments is of great significance, but the diffusion problem of Si during the oxidation process urgently needs to be solved. This chapter analyzes the microstructure and oxidation behavior of Zr–Si amorphous films prepared with different Zr/Si atomic ratios and conducts oxidation experiments at different temperatures. In addition, this chapter also used two other experimental methods: first-principles calculations and molecular dynamics simulations, to study the dominant effects of atomic environment and oxidation-induced diffusion on oxidation behavior. The experimental conclusions of this chapter are as follows:

(a) Microstructure Changes: After oxidation, the film remains amorphous, but forms Zr–Si–O amorphous oxide layer and Si enriched transition amorphous layer.
(b) Chemical Bond Changes: XPS and EELS results indicate that more Zr–Si covalent bonds and Si-O bonds are formed after oxidation, enhancing chemical stability.

Research has found that the oxidation behavior of Zr–Si amorphous thin films varies with different Zr/Si atomic ratios. When high Si content (such as Zr_{55}) oxidizes, a stable amorphous bilayer structure is formed to improve oxidation resistance and chemical stability. When low Si content (such as Zr_{75}) oxidizes, a crystalline oxide layer is formed, causing the oxide layer to crack and peel off. Atomic scale microstructure analysis and first-principles calculations reveal the mechanism of oxidation-induced bond reconstruction and Si self-diffusion-barrier effect, providing insights for the design and optimization of metal Si-based amorphous coating materials. At the same time, Zr–Si amorphous thin films remain amorphous in subcritical water environments, forming Zr–Si covalent bonds and free volume to enhance elastic modulus and plastic deformation ability.

Chapter 5

This chapter explores the oxidation behavior and interfacial barrier effect of Zr–Si multilayer films under high-temperature conditions. The application of multilayer coatings in the protection of nuclear grade zirconium alloys has attracted widespread attention, and nanostructured coatings exhibit excellent performance due to their unique structural evolution. Zr–Si coating is considered an ideal candidate material for protecting zirconium alloy coatings, but its interlayer element diffusion phenomenon at high temperatures has not been systematically studied. Research on zirconium silicon multilayer films at high temperatures has shown that they have excellent antioxidant properties. The amorphous layers formed in hot steam environments have good oxygen-barrier effects and reduce oxygen partial pressure. Atomic scale analysis reveals the selection mechanism and influence of amorphous Zr–Si–O phases, confirming the potential application of Zr/Si multilayer structures as protective coatings for zirconium alloys under nuclear reactor conditions.

(a) Microstructure Changes: After oxidation, the film remains amorphous, but forms Zr–Si–O amorphous oxide layer and Si enriched transition amorphous layer.
(b) Chemical Bond Changes: XPS and EELS results indicate that more Zr–Si covalent bonds and Si-O bonds are formed after oxidation, enhancing chemical stability.
(c) Mechanical Performance Changes: Nanoindentation testing showed that the elastic modulus of the oxidized film increased, the hardness decreased, and the H^3/E^2 ratio decreased, indicating that oxidation enhanced the plastic deformation ability of the film.

Chapter 6

This chapter primarily explores the stability and anticorrosion mechanisms of zirconia (ZrO_2) coatings in the irradiation and oxidation environments of nuclear reactors, aiming to extend the service life of nuclear cladding materials. The research found that the naturally formed zirconia layer on zirconium alloys during reactor operation fails after reaching a certain thickness, mainly due to volumetric expansion from phase transitions, leading to stress and cracks that reduce protective performance. To address this issue, the research team deposited nanocrystalline zirconia coatings on the surface of zirconium alloys and systematically analyzed the effects of plasma excitation power (PLS), target bias (DC), and oxygen partial pressure on the microstructure of the coatings. The results indicate that higher PLS and DC values can increase the content of tetragonal zirconia, thereby enhancing the stability of the coatings.

(a) Failure Mechanism of Zirconia Coatings: The naturally generated zirconia layer fails due to volumetric expansion from phase transitions, resulting in stress and cracks.
(b) Impact of Deposition Process: By adjusting the plasma excitation power and target bias, the proportion of tetragonal zirconia in the coatings can be increased, improving their stability.
(c) Relationship Between Oxygen Defects and Phase Transitions: The study reveals the complex interactions between oxygen defects and the kinetics and thermodynamics of phase transitions.
(d) Impact of Irradiation: The lattice shear induced by Ar^{8+} heavy ion bombardment results in the formation of defect structures in the coatings, enhancing the Zr–O bond strength and improving resistance to irradiation and oxidation.

Chapter 7

ZSO coatings with different Si/Zr ratios were deposited on Zr-4 alloy substrates using magnetron sputtering. Integrating theoretical calculations and experimental data, the study suggests that the $ZrSi_3O$ coating achieves a balance between "diffusion barrier" and "aqueous chemical stability" within the ZSO system, making it an ideal protective coating for zirconium alloys in nuclear reactor environments. Under these optimized conditions, the $ZrSi_3O$ coating exhibits excellent high-temperature and corrosion resistance. However, while higher Si content can enhance oxygen-barrier effects, it may lead to rapid degradation of the coating in high-temperature and high-pressure water environments. Therefore, controlling the Si content is crucial for the coating's application.

(a) Coating Structure and Interface Behavior: When the Zr is dominant in ZSO, both Si and O diffusion occurred at the interface of ZSO/Zr-4 alloy. The surface ZSO layer gradually crystallized into tetragonal zirconia due to the decrease of silicon content, losing its protective ability. When the Si content is high, the coating structure becomes denser, and the electronic structure of Si-O-Zr bonds reduces the diffusion paths for oxygen, significantly inhibiting oxygen penetration and internal Si migration.
(b) Corrosion Performance in High-Temperature Water Environment: The $ZrSi_3O$ coating demonstrates strong stability. Excessive Si content leads to structural degradation and cracking of the coating under prolonged water corrosion due to the high solubility of SiO_2 in water, thus losing its protective function.
(c) High-Temperature Air Oxidation Performance: The $ZrSi_3O$ coating exhibits good oxidation resistance at 400 °C and 600 °C. At 800 °C, although microcracks appear, the coating still provides effective protection, delaying the growth of

the oxidation layer. At 1000 °C, the depth and width of the cracks significantly increase, ultimately leading to coating failure. This may be attributed to the propagation of microcracks and the crystallization of SiO_2 at high temperatures, allowing oxygen ions to diffuse rapidly to the substrate.

Chapter 8

ZrNbCrAlSiO coatings with varying atomic ratios were deposited on Zr-4 alloy substrates to evaluate their protective performance at different temperatures. The structural stability and corrosion resistance of these coatings were assessed through high-temperature, high-pressure hydrothermal corrosion tests and electrochemical corrosion experiments.

(a) High-Temperature Stability of Coatings: ZrNbCrAlSiO coatings with distinct compositions were subjected to high-temperature tests. The $(ZNCA)_1S_3$ coating demonstrated an amorphous nanocrystalline structure after exposure at 1000 °C, which effectively suppressed the growth of nanocrystalline grains. This unique structure provided superior oxygen-blocking properties and mitigated stress accumulation at high temperatures, thereby reducing the likelihood of surface crack formation during high-temperature oxidation.

(b) Protective Effect of $(ZNCA)_1S_3$ Coating on Zr-4 Alloy: Oxidation experiments on Zr-4 alloy at 1000 °C for 2 h revealed that the thickening of the oxide film followed a parabolic growth pattern, regardless of coating presence. However, areas protected by the $(ZNCA)_1S_3$ coating exhibited a significantly slower oxide film growth rate compared to unprotected regions. Additionally, under high-temperature oxidation conditions, the plastic deformation resistance (H^3/E^2) of the $(ZNCA)_1S_3$ coating displayed an increasing trend, albeit remaining lower than that of ZS13 coatings under similar conditions. The lower H^3/E^2 value of the $(ZNCA)_1S_3$ coating indicates enhanced plasticity, which reduces the likelihood of surface cracks.

(c) Corrosion Resistance of Multicomponent Oxide Coatings: The corrosion experiments revealed that $(ZNCA)_1S_3$ coatings are initially unstable in high-temperature, high-pressure LiOH solutions. However, after surface hypoxic modification, M–O–Si mixed bonds (Zr, Nb, Cr, Al) are formed on the surface, which enhances the structural stability of the coating. This modification allows the $(ZNCA)_1S_3$ coating to provide effective protection for the Zr-4 alloy in such environments.

MIX
Papier aus verantwortungsvollen Quellen
Paper from responsible sources
FSC® C105338

If you have any concerns about our products,
you can contact us on
ProductSafety@springernature.com

In case Publisher is established outside the EU,
the EU authorized representative is:
Springer Nature Customer Service Center GmbH
Europaplatz 3, 69115 Heidelberg, Germany

Printed by Libri Plureos GmbH
in Hamburg, Germany